Optimality Conditions

in

Vector Optimization

Editor: Manuel Arana Jiménez (University of Cádiz)

Co-editors: Gabriel Ruiz Garzón (University of Cádiz)
Antonio Rufián Lizana (University of Sevilla)

CONTENT

Foreword

Fuelled by the wealth of applications and by the beauty of the scientific building, optimization has a very long tradition of developments that currently translate into a complex and a very sophisticated body of results. Progress beyond the state-of-the-art requires high quality contributions. This is precisely what this book achieves in the specific area of vector-valued optimization. Here, non-smoothness, various types of generalized convexity, higher order conditions, duality, non-degeneracy, infinite dimensionality of underlying spaces, and non-linearity are compounded with the elaborated solution concepts considered in the area of optimization of vector-valued objective functions.

Let us be more specific.

While in chapter one, it is shown that invexity and pseudo-invexity are appropriate properties for vector-valued functions in that every vector critical point is an efficient or a weakly efficient solution of a Multi-objective Programming Problem, thus generalizing analogue results for scalar case, in chapter three, generalized convexity and variational-like inequalities are used to derive the existence of weakly efficient solutions for some nonsmooth and nonconvex vector optimization problems defined on infinite and finite-dimensional Banach spaces.

Optimality conditions of optimality for nonsmooth continuous-time multiobjective optimization problems under generalized convexity assumptions are derived in both chapters nine and ten. While the former also encompasses problems with smooth data, the later also establishes weak and strong duality theorems for two dual models.

In chapter five, a concept of proper efficiency is presented to develop the optimality conditions and duality results for differentiable non-convex multi-objective programming problems. A relaxation of the proper efficiency assumption enabled the extension of known results to a wider class of nonconvex vector optimization problems. A combination of duality results and necessary and sufficient optimality conditions, now of the Kuhn-Tucker type are obtained in chapter six for non-smooth constrained optimization problems involving generalized type-I functions on the objective and constraint functions involved in the problem.

A constraint qualification condition weaker than those currently available in the literature is derived in chapter two via a unifying approach based on a separation theorem in the vector optimization context.

Chapter three discusses various types of second-order optimality conditions for scalar and vector valued abstract optimization problems. Moreover, second-order sufficient optimality conditions exhibiting a very small gap with the necessary optimality conditions are also considered. The application of these results to derive optimality conditions for optimization problems with equality and inequality constraints is also considered.

Different forms of weak, strong and converse duality theorems are obtained in chapter seven for the Wolfe and Mond-Weir dual problems associated with the vector optimization problem with constraints whose data satisfies assumptions of invexity, strictly invexity or quasi-invexity. Duality results are also derived in chapter eight for pairs of dual programs where the primal is a vector optimization problem with a feasible region defined by a set constraint, equality and inequality constraints, and the duals are of "mixed type" and satisfy suitably generalized concavity properties.

Clearly, this book contains a significant set of contributions which constitute a step forward in shaping the today's Optimization Theory.

Fernando Lobo Pereira
Director of the Institute for Systems and Robotics, Porto
University of Porto

Preface

Vector optimization is continuously needed in several science fields, particularly in economy and engineering. The evolution of these fields depends, in part, on the improvements in vector optimization in mathematical programming.

The search for solutions to vector or multiobjective mathematical programming problems has been carried out through the study of optimality conditions and of the properties of the functions that are involved, as well as through the study of the respective dual problems. In the case of optimality conditions, it is customary to use critical points of the Kuhn-Tucker or Fritz John type. In the case of the classes of functions employed in mathematical programming problems, the tendency has been to substitute convex functions with more general ones, with the objective of obtaining a solution through an optimality condition. Meanwhile the inverse result has also been sought. At the same time, optimality conditions, vector functions and optimality results are being generalized from the differentiable case to the non-differentiable one, and otherwise, extended to other kind of problems, such as continuous time problems.

The aim of this book is to present the last developments in vector optimization. We deeply appreciate the contribution of many of the most relevant researchers in the area that have made possible the existence of this book.

Manuel Arana Jiménez,
Gabriel Ruiz Garzón
and Antonio Rufián Lizana

Contributors

B. Aghezzaf
Département de Mathématiques et d'Informatique, Faculté des Sciences, Université Hassan II Aïn chock, B.P.5366 - Maârif Casablanca Morocco. e-mail: b.aghezzaf@fsac.ac.ma

Tadeusz Antczak
Faculty of Mathematics and Computer Science, University of Lodz, Banacha 22, 90-238 Lodz, Poland. e-mail: antczak@math.uni.lodz.pl

Valeriano Antunes de Oliveira
Universidade Federal de Uberlândia, Faculdade de Ciências Integradas do Pontal. Av. Jose João Dib, 2545 Progresso 38302-000 - Ituiutaba, MG - Brasil. e-mail: valeriano@pontal.ufu.br

M. Arana-Jiménez
Departamento de Estadística e Investigación Operativa., Escuela Superior de Ingeniería, Universidad de Cádiz, 11002 Cádiz , Spain. Tel.: +34 956 015313. e-mail: manuel.arana@uca.es

Lucelina Batista Santos
Departamento de Matematica. Universidade Federal do Paraná. CP 19081 CEP, 81531-990 Curitiba, Paraná, Brazil. e-mail: lucelina@ufpr.br

Adilson J. V. Brandão
Departamento de Matemática. Universidade Federal de São Carlos, Campus Sorocaba. Rodovia João Leme dos Santos, Km 110 - SP-264 Itinga. 18052 780 - Sorocaba, SP - Brasil. e-mail: adilsonvb@ufscar.br

Riccardo Cambini
Department of Statistics and Applied Mathematics, Faculty of Economics, University of Pisa, Via Cosimo Ridolfi 10, 56124 Pisa (Italy). e-mail: cambric@ec.unipi.it

Laura Carosi
Department of Statistics and Applied Mathematics, University of Pisa, Via Ridolfi, 10, 56124 Pisa, Italy. e-mail: lcarosi@ec.unipi.it

M. Hachimi
Université Ibn Zohr, Faculté des Sciences Juridiques Economiques et Sociales, B.P. 8658 Hay Dakhla - Agadir 80000 Morocco. e-mail: hachimi@lycos.com

M. B. Hernández-Jiménez
Departamento de Economía, Métodos Cuantitativos e Ha Económica. Area de

Estadística e Investigación Operativa. Universidad Pablo de Olavide, Spain.
e-mail: mbherjim@upo.es

Laura Martein
Department of Statistics and Applied Mathematics, University of Pisa, Via Ridolfi, 10, 56124 Pisa, Italy. e-mail: lmartein@ec.unipi.it

S. K. Mishra
Department of Mathematics, Faculty of Science, Banaras Hindu University, Varanasi- 221005, India. e-mail: shashikmishra@rediffmail.com

S. Nobakhtian
Department of Mathematics, University of Isfahan, P.O. Box 81745-163, Isfahan, Iran. e-mail: nobakht@math.ui.ac.ir

Sanjay Oli
Department of Mathematics, Asia Pacific Institute of Information Technology SD India. e-mail: sanjayoli@ rediffmail.com

R. Osuna-Gómez
Departamento de Estadística e Investigación Operativa, Facultad de Matemáticas, Universidad de Sevilla, Spain. e-mail: rafaela@us.es

M.R. Pouryayevali
Department of Mathematics, University of Isfahan, P.O. Box 81745-163, Isfahan, Iran. e-mail: pourya@math.ui.ac.ir

J. S. Rautela
Department of Mathematics, Faculty of Applied Science and Humanities, Echelon Institute of Technology, Faridabad 121001, India. e-mail: sky_dreamz@rediffmail.com

Marko Rojas-Medar
Universidad Del Bío-Bío. Facultad de Ciencias. Departamento de Ciencias Básicas. Casilla: 447 Av. Andrés Bello S/N Chillán- Chile. e-mail: marko@ueubiobio.cl

A. Rufián-Lizana
Departamento de Estadística e Investigación Operativa, Facultad de Matemáticas, Universidad de Sevilla, Spain. e-mail: rufian@us.es

G. Ruiz-Garzón
Departamento de Estadística e Investigación Operativa, Facultad de Ciencias Sociales y de la Comunicación, Universidad de Cádiz, Spain. e-mail: gabriel.ruiz@uca.es

L. L. Salles Neto
Departamento de Ciência e Tecnologia, Universidade Federal de São Paulo, Brazil. e-mail: luiz.leduino@unifesp.br

CHAPTER 1

Pseudoinvexity: A good condition for efficiency and weak efficiency in multiobjective mathematical programming. Characterizations.[*]

M. Arana-Jiménez[†]   G. Ruiz-Garzón[‡]   A. Rufián-Lizana[§]

Abstract

We present new classes of vector invex and pseudoinvex functions which generalize the class of scalar invex functions. These new classes of vector functions are characterized in such a way that every vector critical point is an efficient or a weakly efficient solution of a Multiobjective Programming Problem. We establish relationships between these new classes of functions and others used in the study of efficient and weakly efficient solutions, by the introduction of several examples. These results and classes of vector functions are extended to the involved functions in constrained multiobjective mathematical programming problems. It is proved that in order for Kuhn-Tucker points to be efficient or weakly efficient solutions it is necessary and sufficient that the multiobjective problem functions belong to a new class of functions, which we introduce. Similarly, we present characterizations for efficient and weakly efficient solutions by using Fritz John optimality conditions. Some examples are proposed to illustrate these classes of functions and optimality results.

Keywords: Multiobjective programming, invexity, pseudoinvexity, optimality conditions, efficient solutions.

1   Introduction

Nowadays, we can state that optimization in mathematical programming is considered an essential instrument in almost every social, industrial and experimental fields. It is necessary for new applications and theoretical developments, and therefore, new optimization results in mathematical programming are been produced by researchers, which in turn,

---

[*]This work was partially supported by the grant MTM2007-063432 of the Science and Education Spanish Ministry.

[†]Departamento de Estadística e Investigación Operativa., Escuela Superior de Ingeniería, Universidad de Cádiz, 11002 Cádiz , Spain. Tel.: +34 956 015313. e-mail: manuel.arana@uca.es

[‡]Departamento de Estadística e Investigación Operativa, Facultad de Ciencias Sociales y de la Comunicación, Universidad de Cádiz, Spain. e-mail: gabriel.ruiz@uca.es

[§]Departamento de Estadística e Investigación Operativa, Facultad de Matemáticas, Universidad de Sevilla, Spain. e-mail: rufian@uca.es

are been extended to others types of problems. Such as engineering problems, like the control design for autonomous vehicles or impulsive control problems (see [23, 24]), electrical power production (see [8]), economy (see [14]), medicine (see [26]) and ecology (see [17]), among others.

Our objective is to present a brief view on the recent evolution of optimization about the optimality conditions (Fritz John and Kuhn-Tucker types) and the properties of the involved functions (generalized convexity) in a mathematical programming problem. For this purpose, this chapter is based on the recent definitions and results in differentiable multiobjective mathematical programming provided by Arana *et al.*[2, 3, 5, 1, 4], as well as their current research under different conditions. In this way, this chapter is focused on the search for efficient and weakly efficient solutions for mathematical programming problems through the study of optimality conditions and of the properties of the functions that are involved.

In Section 2, we present some important and classical optimality conditions and classes of functions for scalar problems. In this sense, generalized convexity plays an important role. Thanks to the introduction of invex, pseudoinvex [9, 13] and KT-invex [19] classes of functions, some equivalences between critical points and optimal solutions were established; and these classes of functions were characterized. In Section 3, we study the properties of vectorial functions in relation to invexity and pseudoinvextiy. We present the differences between both and their characterizations leading to the conclusion that pseudoinvexity is the adequate property for obtaining efficient and weakly efficient solutions for an unconstrained multiobjective problem. Previous sections are extended in Section 4 and, based on generalized invexity, new classes of vector functions are introduced for the study of efficient and weakly efficient solutions for constrained differenctiable multiobjective programming problems, in such a way that any class of vector functions which is characterized by having every Kuhn-Tucker or Fritz John critical point as an efficient or weakly efficient solution must be equivalent to these classes of functions. Moreover, two examples are proposed to illustrate these new classes of vector functions and some results obtained. Finally, in Section 5, we describe the conclusions.

2   Scalar problem

For scalar problems, in the case of optimality conditions, it is customary to use critical points of the Kuhn-Tucker or Fritz-John [18] types. In the case of the kinds of functions employed in mathematical programming problems, the tendency has been to substitute convex functions with more general ones, with the objective of obtaining a solution through an optimality condition. Meanwhile, the inverse result has sometimes also been sought.

Let us consider the following unconstrained scalar problem:

$$(P) \quad \text{Minimize } \theta(x)$$
$$\text{subject to:}$$
$$x \in S \subseteq \mathbb{R}^n$$

where $\theta : S \subseteq \mathbb{R}^n \to \mathbb{R}$ is a differentiable function on the open set S. It is known that if $\bar{x}$ is a solution, then $\bar{x}$ is a stationary point, that is, $\nabla\theta(\bar{x}) = 0$, where $\nabla\theta(\bar{x})$ is the gradient vector. It is an interesting problem to look for the class of functions for which the reciprocal holds, that is, in which a critical or stationary point $\bar{x}$ is necessarily a solution of (P). In this way, the class of invex functions introduced by Craven and Hanson (see [9, 13]) closed the problem. Let us present some definitions.

Definition 1 *Let $\theta : S \subseteq \mathbb{R}^n \to \mathbb{R}$ be a differentiable function on the open set S. Then, θ is said to be invex if there exists a vector function $\eta : \mathbb{R}^n \times \mathbb{R}^n \to \mathbb{R}^n$ such that $\forall x, \bar{x} \in S$*

$$\theta(x) - \theta(\bar{x}) \geq \nabla\theta(\bar{x})\eta(x, \bar{x}).$$

The class of invex functions was extended to the class of pseudoinvex functions.

Definition 2 *Let $\theta : S \subseteq \mathbb{R}^n \to \mathbb{R}$ be a differentiable function on the open set S. Then, θ is said to be pseudoinvex if there exists a vector function $\eta : \mathbb{R}^n \times \mathbb{R}^n \to \mathbb{R}^n$ such that $\forall x, \bar{x} \in S$*

$$\theta(x) - \theta(\bar{x}) < 0 \Rightarrow \nabla\theta(\bar{x})\eta(x, \bar{x}) < 0.$$

However, these classes of functions are equivalent, (see Ben-Israel and Mond [6]). With the introduction of invex function, the equivalence between a solution of a scalar problem (P) and a stationary point was established; furthermore, this characterizes the invex functions as the next theorem states [6].

Theorem 1 *θ is an invex function if and only if all critical or stationary points are optimal solutions of (P).*

That is,

$$\boxed{\theta \text{ INVEX}}$$

$$\Updownarrow$$

$$\boxed{\bar{x} \text{ Stationaty point} \Longleftrightarrow \bar{x} \text{ Optimal solution}}$$

Several authors have generalized convexity and invexity, and they have continued the study of optimal solutions for constrained scalar problems ([11], [15], [25],...), formulated as follows:

$$(CP) \quad \text{Minimize } \theta(x)$$
$$\text{subject to:}$$
$$g(x) \leqq 0$$

$$x \in S \subseteq R^n$$

where $\theta : S \subseteq R^n \to R$, $g = (g_1, \ldots, g_m) : S \subseteq R^n \to R^m$, $\theta, g_1, \ldots, g_m$ are differentiable functions on the open set $S \subseteq R^n$.

For constrained scalar problems (CP), invexity is a sufficient condition for which a Kuhn-Tucker critical point leads to a solution of (CP), but it is not necessary. Martin [19] defined a weaker concept, called KT-invexity, and proved that it is a necessary and sufficient condition in order for a Kuhn-Tucker critical point to be an optimal solution of (CP).

Definition 3 *Problem (CP) is said to be KT-invex if there exists a function $\eta : S \times S \to R^n$ such that $\forall x, \bar{x} \in S$, with $g_i(x) \leq 0, g_i(\bar{x}) \leq 0$, $i = 1, \ldots, m$,*

$$\theta(x) - \theta(\bar{x}) > \nabla\theta(\bar{x})\eta(x, \bar{x}),$$

$$-\nabla g_j(\bar{x})\eta(x, \bar{x}) \geq 0, \qquad \forall j \in I(\bar{x}),$$

where

$$I(\bar{x}) = \{j : j = 1, \ldots, m \text{ such that } g_j(\bar{x}) = 0\}.$$

Martin [19] obtained the following result:

Theorem 2 *Every Kuhn-Tucker critical point is an optimal solution of (CP) if and only if (CP) is KT-invex.*

Therefore, KT-invexity is characterized, such as the following graph shows:

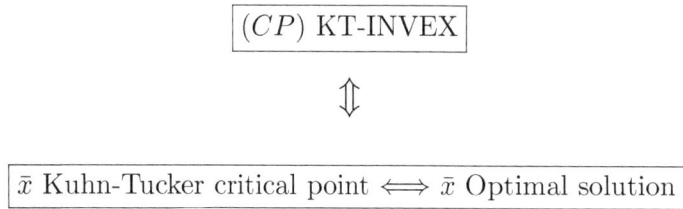

$$\boxed{(CP) \text{ KT-INVEX}}$$

$$\Updownarrow$$

$$\boxed{\bar{x} \text{ Kuhn-Tucker critical point} \iff \bar{x} \text{ Optimal solution}}$$

3 Unconstrained multiobjective problem

Let us introduce the following convention for equalities and inequalities.
If $x = (x_1, \ldots, x_n)$, $y = (y_1, \ldots, y_n) \in \mathbb{R}^n$, then

$$
\begin{aligned}
x = y &\Leftrightarrow x_i = y_i, &\forall i = 1, \ldots, n, \\
x < y &\Leftrightarrow x_i < y_i, &\forall i = 1, \ldots, n, \\
x \leqq y &\Leftrightarrow x_i \leq y_i, &\forall i = 1, \ldots, n, \\
x \leq y &\Leftrightarrow x_i \leq y_i, &\forall i = 1, \ldots, n, &\text{and there exists } j \text{ such that } x_j < y_j.
\end{aligned}
$$

Similarly, $>, \geq, \geq$. Firstly, let us center on unconstrained multiobjective problems, which, in general, can be formulated as follows:

$$(MP) \quad \text{Minimize } f(x) = (f_1(x), \ldots, f_p(x))$$
$$\text{subject to:}$$

$$x \in S \subseteq \mathbb{R}^n$$

where $f : S \subseteq \mathbb{R}^n \to \mathbb{R}^p$ is differentiable and S is said to be the feasible set. The solutions of (MP) are named efficient points and they were introduced by Pareto [22].

Definition 4 *A feasible point, $\bar{x}$, is said to be an efficient solution of (MP) if there does not exist another feasible point, x, such that $f(x) \leq f(\bar{x})$.*

Later, it appeared a more general concept such as the weakly efficient solution of (MP).

Definition 5 *A feasible point, $\bar{x}$, is said to be a weakly efficient solution of (MP) if there does not exist another feasible point, x, such that $f(x) < f(\bar{x})$.*

It is easy to see that any efficient point is a weakly efficient point. The following optimality condition type is usually employed.

Definition 6 *A feasible point for (MP), $\bar{x}$, is said to be a vector critical point if there exists $\lambda \in \mathbb{R}^p$, such that*

$$\lambda^T \nabla f(\bar{x}) = 0,$$

$$\lambda \geq 0,$$

where $\nabla f(\bar{x}) \in \mathbb{M}^{p \times n}$ is the gradient matrix of the vector function f, and $\mathbb{M}^{p \times n}$ denotes the set of $p \times n$ real matrices. The vector critical point condition is necessary for a feasible point of (MP) to be an efficient solution (see [16, 12]) or a weakly efficient solution (see [10]), as the next theorem states.

Theorem 3 *If $\bar{x}$ is a weakly efficient solution for (MP), then $\bar{x}$ is a vector critical point.*

Now, the objective is to extend the result in Theorem 1 to multiobjective programming. For that purpose, let us consider the following definition, which extends the concept of scalar invex function to the multiple case (see [20, 21]).

Definition 7 *Let $f = (f_1, \ldots, f_p) : S \subseteq \mathbb{R}^n \to \mathbb{R}^p$ be a differentiable function on the open set S. Then, the vector function f is said to be invex if there exists a vector function $\eta : \mathbb{R}^n \times \mathbb{R}^n \to \mathbb{R}^n$ such that $\forall x, \bar{x} \in S$*

$$f(x) - f(\bar{x}) \geq \nabla f(\bar{x}) \eta(x, \bar{x})$$

A vector invex function is characterized by its scalar components, as follows:

Proposition 1 *Let $f = (f_1, \ldots, f_p) : S \subseteq \mathbb{R}^n \rightarrow \mathbb{R}^p$ be a differentiable function on the open set S. Then f is invex with respect to the vector function η if and only if the scalar functions f_i, $1 \leq i \leq p$, are invex with respect to the same vector function η.*

Since the previous invex definition generalizes Definition 1, we could think that their properties are generalized to the vectorial case too, that is, for example Theorem 1 would occur for efficient or weakly efficient solutions. But this is not the case as it is showed in the following example.

Example 1 *Let us consider $f = (f_1, f_2) : \mathbb{R} \rightarrow \mathbb{R}^2$ with $f(x) = (f_1(x), f_2(x)) = (x^2, -x^2)$.*

$$(P1) \quad Minimize \; f(x) = (f_1(x), \ldots, f_p(x))$$
$$subject \; to:$$

$$x \in \mathbb{R}^n$$

$\forall \; x, \bar{x} \in \mathbb{R}$ we have that $x^2 - \bar{x}^2 < 0 \Leftrightarrow -x^2 + \bar{x}^2 > 0$, i.e.,

$$f_1(x) - f_1(\bar{x}) < 0 \Leftrightarrow f_2(x) - f_2(\bar{x}) > 0.$$

Hence, $\nexists \; x, \bar{x} \in \mathbb{R}$ such that $f_i(x) - f_i(\bar{x}) < 0$, $i = 1, 2$, which implies that $f(x) - f(\bar{x}) < 0$ is not verified. In the same way, $f(x) - f(\bar{x}) \leq 0$ is not fulfilled. Therefore, every point is both an efficient and weakly efficient solution. So, we can state that every vector critical point is an efficient and weakly efficient solution for P.

On the other hand, f_2 is not invex because $\nabla f_2(0) = 0$ and $\bar{x} = 0$ is not a minimum for this function. By Proposition 1, we have that f is not invex. Consequently, we conclude that invexity is not a necessary condition for every vector critical point to be efficient or weakly efficient solution for (P1).

Example 2 *Let us consider:*

$$(P2) \quad Minimize \; f(x) = (f_1(x), \ldots, f_p(x))$$
$$subject \; to:$$

$$x \in \mathbb{R}^n$$

where $f = (f_1, f_2) : \mathbb{R} \rightarrow \mathbb{R}^2$ with $f(x) = (f_1(x), f_2(x)) = (x^2, 5)$. We have that $\nabla f(x) = (2x, 0)$. $\forall x, \bar{x} \in \mathbb{R}$, let us see that f is invex with respect to the vector function

$$\eta(x, \bar{x}) = \begin{cases} \dfrac{x^2 - \bar{x}^2}{2\bar{x}} & if \quad \bar{x} \neq 0, \\ 1 & if \quad \bar{x} = 0. \end{cases}$$

It follows that

$$f(x) - f(\bar{x}) = (x^2 - \bar{x}^2, 0) = \left\{ \begin{array}{ll} (\dfrac{x^2 - \bar{x}^2}{2\bar{x}} \cdot 2\bar{x}, 0) & if \quad \bar{x} \neq 0 \\ (x^2, 0) & if \quad \bar{x} = 0 \end{array} \right\} \geqq \nabla f(\bar{x}) \eta(x, \bar{x}).$$

In consequence, f is invex. Let us take $\bar{x} = 7$ and $\lambda = (0, 4)$, it follows $\lambda^T \nabla f(7) = 0$, and then $\bar{x} = 7$ is a vector critical point. However, it is not an efficient solution for (P2), since given $x = 5$, $f(5) - f(7) = (-24, 0) \leq (0, 0)$

Therefore, invexity is not the answer to our problem. New vector function properties were necessary to obtain efficient and weakly efficient solutions from vector critical point. For this purpose, Osuna *et al.*[20], and recently Arana *et al.*[2] define two classes of vector functions generalizing the class of scalar pseudoinvex functions.

Definition 8 *Let $f = (f_1, \ldots, f_p) : S \subseteq \mathbb{R}^n \to \mathbb{R}^p$ be a differentiable function on the open set S. Then, the vector function f is said to be pseudoinvex-I if there exists a vector function $\eta : \mathbb{R}^n \times \mathbb{R}^n \to \mathbb{R}^n$ such that $\forall x, \bar{x} \in S$*

$$f(x) - f(\bar{x}) < 0 \Rightarrow \nabla f(\bar{x})\eta(x, \bar{x}) < 0.$$

Definition 9 *Let $f = (f_1, \ldots, f_p) : S \subseteq \mathbb{R}^n \to \mathbb{R}^p$ be a differentiable function on the open set S. Then, the vector function f is said to be pseudoinvex-II if there exists a vector function $\eta : \mathbb{R}^n \times \mathbb{R}^n \to \mathbb{R}^n$ such that $\forall x, \bar{x} \in S$*

$$f(x) - f(\bar{x}) \leq 0 \Rightarrow \nabla f(\bar{x})\eta(x, \bar{x}) < 0.$$

In the scalar case, i.e, when $p = 1$, Definitions 7, 8 and 9 are equivalent (see [6]).

The vector critical point condition is a necessary optimality condition for a point to be an efficient or weakly efficient solution of the multiobjective problem (MP), (see [13, 10]). Osuna *et al.* [20] proved that pseudoinvexity-I is necessary and sufficient for the set of vector critical points and the set of weakly efficient solutions of (MP) to be equivalent.

Theorem 4 *Every vector critical point is a weakly efficient solution of (MP) if and only if f is pseudoinvex-I.*

$$\boxed{f \text{ PSEUDOINVEX-I}}$$

$$\Updownarrow$$

$$\boxed{\bar{x} \text{ Vector critical point} \iff \bar{x} \text{ Weakly efficient solution}}$$

Recently, Arana *et al.*[2] have extended this result to the study of efficient solutions, as follows:

Theorem 5 *Every critical point is an efficient solution of (MP) if and only if f is pseudoinvex-II.*

This result generalizes Theorem 1 to the multiple case. Furthermore, the equivalence between vector critical points and efficient solutions of (MP) characterizes the class of pseudoinvex-II vector functions.

Therefore, the equivalence between vector critical points and efficient solutions of (MP) characterizes the class of pseudoinvex-II vector functions such as follows.

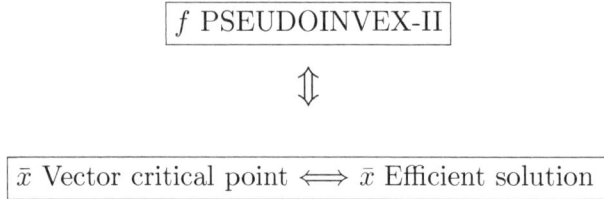

$$\boxed{f \text{ PSEUDOINVEX-II}}$$

$$\Updownarrow$$

$$\boxed{\bar{x} \text{ Vector critical point} \iff \bar{x} \text{ Efficient solution}}$$

We have seen that pseudoinvexity plays an important role in (MP). In the scalar case, the classes of invex, pseudoinvex-I and pseudoinvex-II functions are equivalent. However, these are different classes of functions in the vectorial case, such as the following theorem establishes [2].

$PSI = \{f : S \subseteq R^n \to R^p \ / \ f \text{ is pseudoinvex-I}\}$,
$PSII = \{f : S \subseteq R^n \to R^p \ / \ f \text{ is pseudoinvex-II}\}$,
$INV = \{f : S \subseteq R^n \to R^p \ / \ f \text{ is invex}\}$.

Theorem 6 $INV \cup PSII \subset PSI$ *and* $INV \cup PSII \neq PSI$

Consequently, the relationship between invex, pseudoinvex-I and pseudoinvex-II functions is as follows:

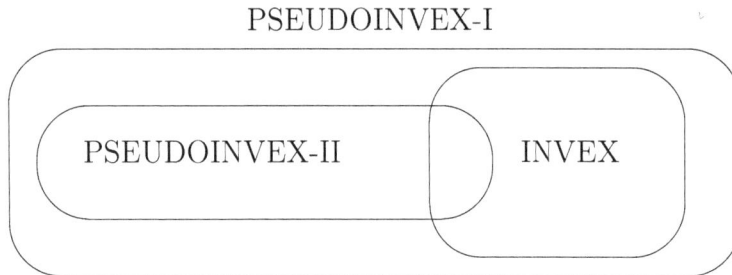

PSEUDOINVEX-I

4   Constrained multiobjective problem

Now, let us add some constraints to our multiobjective problem:

$$(MP) \quad \text{Minimize } f(x) = (f_1(x), \ldots, f_p(x))$$
$$\text{subject to:}$$

$$x \in S \subseteq \mathbb{R}^n$$

where $g = (g_1, \ldots, g_m) : S \subseteq \mathbb{R}^n \to \mathbb{R}^m$ is differentiable. Let denote K the feasible set. In a similar way as for (CP), Fritz John and Kuhn-Tucker conditions allow us to obtain efficient and weakly efficient solutions for (CMP).

Definition 10 *A feasible point $\bar{x}$ for (CMP) is said to be a Fritz-John vector critical point, FJVCP, if there exist $\lambda \in \mathbb{R}^p$, $\mu \in \mathbb{R}^m$ such that*

$$\lambda^T \nabla f(\bar{x}) + \mu^T \nabla g(\bar{x}) = 0 \tag{1}$$

$$\mu^T g(\bar{x}) = 0 \tag{2}$$

$$(\lambda, \mu) \geq 0, \quad (\lambda, \mu) \neq 0 \tag{3}$$

Definition 11 *A feasible point $\bar{x}$ for (CMP) is said to be a Kuhn-Tucker vector critical point, KTVCP, if there exist $\lambda \in \mathbb{R}^p$, $\mu \in \mathbb{R}^m$ such that*

$$\lambda^T \nabla f(\bar{x}) + \mu^T \nabla g(\bar{x}) = 0 \tag{4}$$

$$\mu^T g(\bar{x}) = 0 \tag{5}$$

$$\mu \geq 0 \tag{6}$$

$$\lambda \geq 0 \tag{7}$$

It is easy to see that every KTVCP is FJVCP. Based on results by Chankong and Haimes [7], Kanniappan [16], and Gulati and Talaat [12], the following Fritz John and Kuhn-Tucker optimality result is obtained for efficient and weakly efficient solutions of (CMP).

Theorem 7 *If $\bar{x}$ is an efficient or weakly efficient solution of (CMP), then $\bar{x}$ is a FJVCP.*

Now, we need to take on a constraint qualification.

Theorem 8 *If $\bar{x}$ is an efficient or weakly efficient solution of (CMP) and a constraint qualification is satisfied at $\bar{x}$, then $\bar{x}$ is a KTVCP.*

The Fritz John and Kuhn-Tucker optimality conditions are necessary for a point to be an efficient solution for (CMP), as we have already seen, but these conditions are not necessary in general. Many properties based on generalized convexity have been proposed in order to state that a Fritz John or Kuhn-Tucker point is an efficient or weakly efficient solution.

Osuna *et al.* [21] proved that for weakly efficient solutions through Kuhn-Tucker optimality conditions, this search is resolved by the KT-pseudoinvex-I class, which we defined as follows:

Definition 12 *The problem (CMP) is said to be KT-pseudoinvex-I if there exists a vector function $\eta : S \times S \to \mathbb{R}^n$ such that for all feasible points $x, \bar{x}$*

$$f(x) - f(\bar{x}) < 0 \Rightarrow \begin{cases} \nabla f(\bar{x})\eta(x, \bar{x}) < 0 \\ \nabla g_j(\bar{x})\eta(x, \bar{x}) \leqq 0, \quad \forall j \in I(\bar{x}) \end{cases}$$

where $I(\bar{x}) = \{j = 1, \ldots, m : g_j(\bar{x}) = 0\}$.

And Osuna *et al.* [21] established the following characterization theorem for (CMP).

Theorem 9 *Every KTVCP is a weakly efficient solution of (CMP) if and only if problem (CMP) is KT-pseudoinvex-I.*

Recently, Arana *et al.*[3] have extended this study to the localization of weakly efficient solutions from Fritz John optimality conditions. For that, they have introduced the following concept.

Definition 13 *The problem (CMP) is said to be FJ-pseudoinvex-I if there exists a vector function $\eta : S \times S \to \mathbb{R}^n$ such that for all feasible points $x, \bar{x}$*

$$f(x) - f(\bar{x}) < 0 \Rightarrow \begin{cases} \nabla f(\bar{x})\eta(x, \bar{x}) < 0 \\ \nabla g_j(\bar{x})\eta(x, \bar{x}) < 0, \quad \forall j \in I(\bar{x}) \end{cases}$$

where $I(\bar{x}) = \{j = 1, \ldots, m : g_j(\bar{x}) = 0\}$.

And proceeding as in Osuna *et al.*[21], Arana *et al.*[3] establish and prove the following characterization theorem for (CMP).

Theorem 10 *Every FJVCP is a weakly efficient solution of (CMP) if and only if problem (CMP) is FJ-pseudoinvex-I.*

However, finding a class of functions for which it was possible to verify not only that a Kuhn-Tucker or Fritz John vector critical point is an efficient solution, but it was also characterized by this, was still left pending. Thanks to KT and FJ-pseudoinvexity, Arana *et al.*[3] have recently closed this problem.

Definition 14 *Problem (CMP) is said to be KT-pseudoinvex-II if there exists a vector function $\eta : S \times S \to \mathbb{R}^n$ such that for all feasible points $x, \bar{x}$*

$$f(x) - f(\bar{x}) \leq 0 \Rightarrow \begin{cases} \nabla f(\bar{x})\eta(x, \bar{x}) < 0 \\ \nabla g_j(\bar{x})\eta(x, \bar{x}) \leq 0, \quad \forall j \in I(\bar{x}) \end{cases}$$

where $I(\bar{x}) = \{j = 1, \ldots, m : g_j(\bar{x}) = 0\}$.

Definition 15 *Problem (CMP) is said to be FJ-pseudoinvex-II if there exists a vector function $\eta : S \times S \to \mathbb{R}^n$ such that for all feasible points $x, \bar{x}$*

$$ f(x) - f(\bar{x}) \leq 0 \Rightarrow \begin{cases} \nabla f(\bar{x}) \eta(x, \bar{x}) < 0 \\ \nabla g_j(\bar{x}) \eta(x, \bar{x}) < 0, \quad \forall j \in I(\bar{x}) \end{cases} $$

where $I(\bar{x}) = \{j = 1, \ldots, m : g_j(\bar{x}) = 0\}$.

Arana *et al.*[3] have proved that KT and FJ-pseudoinvexity-II are both necessary and sufficient for a Kuhn-Tucker point and a Fritz John point to be an efficient solution for (CMP), respectively, as follows:

Theorem 11 *Every KTVCP is an efficient solution of (CMP) if and only if (CMP) is KT-pseudoinvex-II.*

Similarly, we have the following characterization result:

Theorem 12 *Every FJVCP is an efficient solution of (CMP) if and only if (CMP) is FJ-pseudoinvex-II.*

These results for efficient solutions of (CMP) can be considered a generalization of results for the scalar problem (CP) given by Martin [19]. Furthermore, this result characterizes the KT and FJ-pseudoinvex-II multiobjective problems. The following scheme shows these results:

$$ \boxed{\text{(CMP) KT-Pseudoinvex-I}} \iff \boxed{\bar{x} \text{ KTVCP} \iff \bar{x} \text{ Weakly efficient solution}} $$

$$ \boxed{\text{(CMP) FJ-Pseudoinvex-I}} \iff \boxed{\bar{x} \text{ FJVCP} \iff \bar{x} \text{ Weakly efficient solution}} $$

$$ \boxed{\text{(CMP) KT-Pseudoinvex-II}} \iff \boxed{\bar{x} \text{ KTVCP} \iff \bar{x} \text{ Efficient solution}} $$

$$ \boxed{\text{(CMP) FJ-Pseudoinvex-II}} \iff \boxed{\bar{x} \text{ FJVCP} \iff \bar{x} \text{ Efficient solution}} $$

In the same way as in the previous section, we have that the properties of the vectorial invexity and the vectorial psedoinvexity cases are different. In this sense, we show that invexity is not a necessary condition for every vector critical point to be efficient o weakly efficient solution, and even more, that invexity is not a sufficient condition for it. For that purpose, let us consider the following examples (see [3]) to illustrate these situations and the optimality results related to KT-pseudoinvexity and invexity.

$(x,f(x))=(x,x^{**}2+1,-x^{**}2+2)$ ———

z=-x**2+2

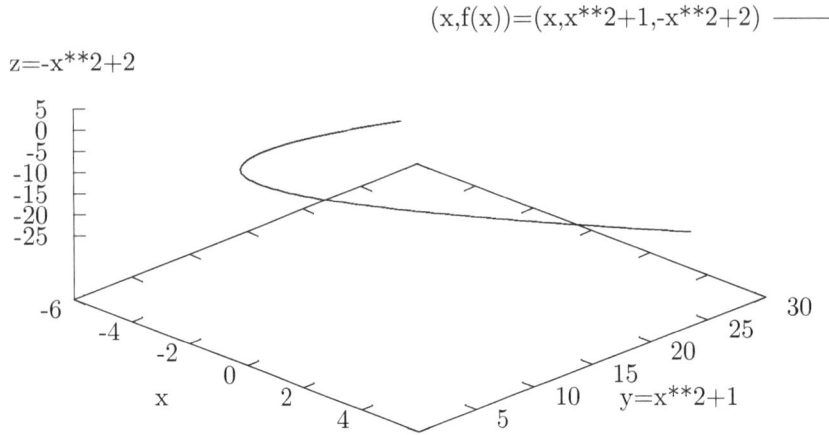

Figure 1: objective function graph for (CMP1)

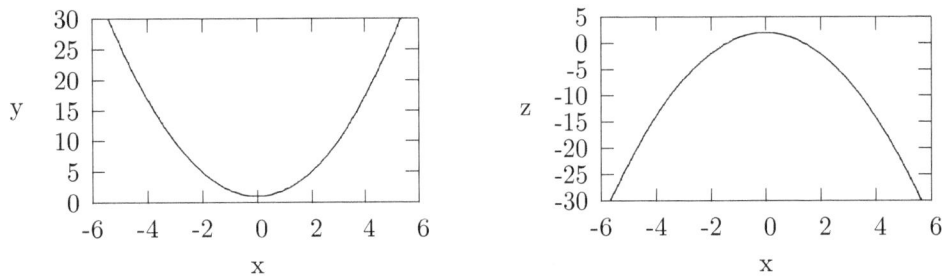

Figure 2: projections of objective function graph for (CMP1)

Example 3 *We present an example of KT-pseudoinvex-II multiobjective problem, in which we show that all KTVCP are efficient solutions. Besides, we prove that invexity of the functions involved in this problem, f and g, is not verified.*

$$(CMP1) \qquad Minimize \ (x^2 + 1, -x^2 + 2)$$
$$subject \ to:$$

$$3 - x \leq 0$$

$$x \in R$$

where $f = (f_1, f_2) : \mathbb{R} \to \mathbb{R}^2$, with $f(x) = (f_1(x), f_2(x)) = (x^2+1, -x^2+2)$, and $g : \mathbb{R} \to \mathbb{R}$, with $g(x) = 3 - x$, are differentiable functions on $\mathbb{R}$. An objective function graph is shown in Figure 1, and its projections on XY and XZ planes are also shown in Figure 2 to *provide a better.*

It is verified:

(i) (CMP1) is KT-pseudoinvex-II.

(ii) f is not invex.

(iii) All feasible points are KTVCP.

(iv) All feasible points are efficient solutions.

Consequently, (CMP1) is a KT-pseudoinvex-II problem, where all KTVCP are efficient solutions, and however the objective function f is not invex.

Example 4 *Let us consider a new problem (CMP2) with f and g invex functions, which is not KT-pseudoinvex-II. We will provide some KTVCP that is not an efficient solution.*

$$(CMP2) \qquad Minimize \ (x^2 - 3, 10)$$
$$subject \ to:$$
$$-1 - x \le 0$$
$$x \in \mathbb{R}$$

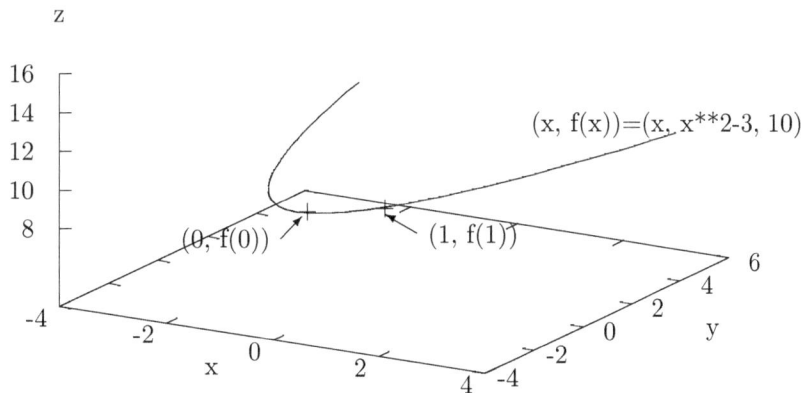

Figure 3: objective function graph for (CMP2)

where $f = (f_1, f_2) : \mathbb{R} \to \mathbb{R}^2$, with $f(x) = (f_1(x), f_2(x)) = (x^2 - 3, 10)$, and $g : \mathbb{R} \to \mathbb{R}$, with $g(x) = -1 - x$, are differentiable functions on $\mathbb{R}$. Figure 3 shows the objective function graph.

The following is verified:

(i) f and g are invex.

(ii) (CMP2) is not KT-pseudoinvex-II.

(iii) Let us consider the feasible point $\bar{x} = 1$. The point $\bar{x}$ is a KTVCP. Otherwise, $\bar{x}$ is not an efficient solution for (CMP2), since given the feasible point $x = 0$, we have that $f(x) - f(\bar{x}) = (-1, 0) \leq 0$. They are both illustrated in Figure 3.

So, there exists a KTVCP, which is not an efficient solution for the multiobjective problem (CMP2), and however f and g are invex functions.

5   Conclusion

In this chapter we have show the evolution of the generalized convexity in mathematical programming problems for the localization of solutions from optimality conditions. We have seen the generalization of definitions and results of scalar problems to multiobjective problems, and these in turn, adequated to some types of solutions. In this sense, we can tackle the search for efficient and weakly efficient solutions from vector optimality conditions of the Kuhn-Tucker and Fritz John types. For the Kuhn-Tucker case, we have introduced two kinds of vector functions, the KT-pseudoinvex-I and II, which are characterized by every Kuhn-Tucker point being a weakly efficient and an efficient solution, respectively. Analogously, the Fritz-John optimality condition is a necessary and sufficient condition for a point to be a weakly efficient or an efficient solution beneath FJ-pseudoinvexity-I or II, respectively, and moreover these kinds of vector functions are characterized by this property. Therefore, we have generalized the characterizations about optimality provided by Martin [19] for scalar problems. We have also illustrated these optimality results with some examples.

References

[1] M. Arana-Jiménez, Efficiency under pseudoinvexity and duality in differentiable and non-differentiable multiobjective problems: a characterization, in Recent development in mathematical programming, Aracne editrice S.r.l., Roma, 2009.

[2] M. Arana-Jiménez, A. Rufián-Lizana, R. Osuna-Gómez and G. Ruiz-Garzón, A characterization of pseudoinvexity in multiobjective programming, Mathematical and Computer Modelling 48 (2008) 1719-1723.

[3] M. Arana-Jiménez, A. Rufián-Lizana, R. Osuna-Gómez and G. Ruiz-Garzón, Pseudoinvexity, optimality conditions and efficiency in multiobjective problems; duality, Nonlinear Analysis 68 (2008) 24-34.

[4] M. Arana-Jiménez, G. Ruiz-Garzón, R. Osuna-Gómez and A. Rufián-Lizana, Efficiency under pseudoinvexity in nonsmooth multiobjective problems, Book of abstracts of EURO Conference 23, Bonn, 2009.

[5] M. Arana-Jiménez, G. Ruiz-Garzón, A. Rufián-Lizana and R. Osuna-Gómez, Invexity and pseudoinvity in multiobjective programming, Proceeding of SEIO XXXI, Murcia, 2009.

[6] A. Ben-Israel and B. Mond, What is Invexity, Journal of the Australiam Mathematical Society, Series B 28 (1986) 1-9.

[7] V. Chankong and Y.Y. Haimes, Multiobjective Decision Making: Theory and Methodology, North-Holland, New York, 1983.

[8] G.S. Christensen, M.E. El-Hawary, S.A. Soliman, Optimal control applications in electric power system, Plenum, New York, 1987.

[9] B.D. Craven, Invex Functions and Constraint Local Minima, Journal of the Australiam Mathematical Society, Series B 24 (1981) 357-366.

[10] B.D Craven, Lagrangian Conditions and Quasiduality, Journal of the Australiam Mathematical Society, Series B 16 (1977) 325-339.

[11] R.R. Egudo and M.A. Hanson, Duality with Generalized Convexity, Journal of the Australiam Mathematical Society, Series B 28 (1986) 10-21.

[12] T.R. Gulati and N. Talaat, Sufficiency and Duality in Nondifferentiable Multiobjective Programming, Opsearch 28, 2 (1991) 73-87.

[13] M.A. Hanson, On Sufficiency of Kuhn-Tucker Conditions, Journal of Mathematical Analysis and Applications 80 (1981) 545-550.

[14] M.D. Intriligator, Mathematical optimization and economic theory, Prentice-Hall, Englewood Cliffs, New Jersey, 1971.

[15] Y. Jeyakumar, Strong and Weak Invexity in Mathematical Programming, Methods of Operations Research 55 (1990) 109-125.

[16] P. Kanniappan, Necessary Conditions for Optimality of Nondifferentiable Convex Multiobjective Programming, Journal of Optimization Theory and Applications 40 (1983) 167-174.

[17] G. Leitmann, The calculus of variations and optimal control, Plenum Press, New York, 1981.

[18] O.L. Mangasarian, Nonlinear Programming, McGraw Hill Book Company, New York, 1969.

[19] D.M. Martin , The Essence of Invexity, Journal of Optimization Theory and Applications 47, 1 (1985) 65-76.

[20] R. Osuna-Gómez, A. Rufián-Lizana and P. Ruiz-Canales, Invex Functions and Generalized Convexity in Multiobjective Programming, Journal of Optimization Theory and Applications 98 (1998) 651-661.

[21] R. Osuna-Gómez, A. Beato-Moreno and A. Rufián-Lizana, Generalized Convexity in Multiobjective Programming, Journal of Mathematical Analysis and Applications 233 (1999) 205-220.

[22] V. Pareto, Course d'economie politique, Rouge, Lausanne, 1896.

[23] F.L. Pereira, Control design for autonomous vehicles: a dynamic optimization perspective, European Journal of Control 7 (2001) 178-202.

[24] F.L. Pereira, A maximum principle for impulsive control problems with state constraints, Computational and Applied Mathematics 19 (2000) 1-19.

[25] V. Preda, On Duality with Generalized Convexity, Bollettino della Unione Matematica Italiana, Sezione A (7) (1991) 5.

[26] G.W. Swam, Applications of optimal control theory in biomedicine, Marcel Dekker, New York, 1984.

Optimality Conditions in Vector Optimization, 2010, 17-34

CHAPTER 2

Optimality and constraint qualifications in vector optimization

Carosi Laura and Martein Laura *

Abstract

We propose a unifying approach in deriving constraint qualifications and theorem of the alternative. We first introduce a separation theorem between a subspace and the non-positive orthant, and then we use it to derive a new constraint qualification for a smooth vector optimization problem with inequality constraints. The proposed condition is weaker than the existing conditions stated in the recent literature. According with the strict relationship between generalized convexity and constraint qualifications, we introduce a new class of generalized convex vector functions. This allows us to obtain some new constraint qualifications in a more general form than the ones related to componentwise generalized convexity. Finally, the introduced separation theorem allows us to derive some of the known theorems of the alternative which are used in the literature to get constraint qualifications.

Keywords: Constraint qualifications, generalized convex vector functions, multi-objective programming, optimality conditions.

1 Introduction

As it is well known, constraint qualifications play a fundamental role for both a scalar and a vector optimization problem, since they guarantee the validity of the Karush-Kuhn-Tucker necessary optimality conditions, i.e., the positivity of the multiplier associated with the objective functions in Fritz John conditions.

Since the early fifties, several constraint qualification conditions have been established for scalar optimization problems either in a smooth or in a non-smooth context, and, regarding the feasible region, inequality and equality and set constraints have been dealt with[1]. All the proposed conditions impose some regularity on the behavior of the feasible region. Among them, the so called Guignard constraint qualification is considered as the weakest one in the sense that all the other constraint qualifications imply it. Among the most famous constraint qualifications we recall Abadie, Cottle, Slater, Mangasarian-Fromovitz

---

*Department of Statistics and Applied Mathematics, University of Pisa, Via Ridolfi, 10, 56124 Pisa, Italy, email: lcarosi@ec,unipi.it, lmartein@ec.unipi.it

[1]For a general presentation of constraint qualification for scalar optimization problems, the reader can refer to [1] and references therein.

constraint qualifications and linear objective constraint qualification.

As in the case of sufficient optimality conditions, the study of constraint qualifications has benefited from the developments on generalized convexity, so that several conditions have been established by imposing some suitable generalized convexity properties on the constraint functions.

The key tools in studying constraint qualifications are the theorems of the alternative, in the sense that the status of "qualification constraints" related to a given condition is, in general, proved by means of a suitable theorem of the alternative.

The classical analysis of constraint qualifications for a scalar problem has been extended to multiobjective programs by Maeda in [14]. In this path-breaking paper, Maeda considers a vector optimization problem where the objective and the constraint functions are differentiable and the feasible region is defined by inequality constraints. He observes that, unlike the scalar case, constraint qualifications must concern the behavior of both the objective functions and the feasible region. For this reason, in vector optimization, some authors prefer referring to constraint qualification as regularity conditions (see for instance [7, 16]). Meada introduces the so called Generalized Guignard constraint qualfication which is, according to his scheme, the weakest condition and then, he extends to the vector case other relevant scalar conditions.

The analysis, started with Maeda's paper, has been developed through different directions. For instance, Preda and Chitescu [19] study constraint qualifications for a directionally differentiable vector optimization problem having a feasible region defined by inequality constraints. In Giorgi *et al.* [8] both inequality and equality constraints are considered and, as in [19], all the involved functions are directionally differentiable. In the same directionally differentiable context, Nguyen and Luu [18] consider a feasible region defined by inequality, equality and set constraint. The case of locally Lipschitz functions is developed by Chandra *et al.* [7], Li [11], and Li and Zhang [12].

As some mentioned papers underline, like in the scalar case, generalized convexity plays a key role in deriving new constraint qualifications for vector optimization problems (see for instance [7, 8, 9, 11, 12, 19]).

Furthermore, we would like to point out that, even for vector optimization problems, there exists a strict relationship between constraint qualifications and theorems of the alternative: all the obtained results are proved by means of suitable theorems of the alternative and some new constraint qualifications have been introduced by providing new theorems of the alternative.

In this chapter, we will propose a unifying approach in deriving constraint qualifications and theorems of the alternative. More exactly, we will state and prove a separation theorem between a subspace and the non-positive orthant, which establishes the sign of multipliers in connection with the intersection of the subspace with the boundary of the non-positive orthant. The obtained result will allow us to state a necessary and sufficient condition for having regularity in a differentiable vector optimization problem with inequality constraints, i.e., a condition which characterizes the positivity of the multiplier vector associated with the objective functions in Fritz John conditions. We will prove that the Generalized Guignard constraint qualification, introduced by Maeda [14], implies our necessary and sufficient condition and we will provide an example to put in evidence that the reverse implication does not hold. Therefore, the existing constraint qualifications presented in the literature for smooth vector optimization problems with inequality constraints follow from our characterization. It is worth noticing that we do not use a theorem of the alternative for establishing our condition. On the contrary, we will put in evidence how our separation theorem allows us to obtain a new approach in deriving theorems of the alternative.

Finally, along the line of the recent literature, we will study the role of generalized convexity in connection with constraint qualifications. In this light, we will introduce a new class of generalized convex vector functions in order to obtain some constraint qualifications which are more general than the ones requiring componentwise generalized convexity (see for istance [8, 11, 12, 19]).

The chapter is organized as follows: in Section 2, we will state the separation theorem, while Section 3 and Section 4 will be devoted to the study of constraint qualifications; the relationship between the new conditions and the existing ones is also deepened. In Section 5, we will show how the proposed separation theorem allows us to derive some well known theorems of the alternative.

Throughout this chapter we will use the following definitions and notations.

Definition 1 *Let X be a set and let $z \in clX$. The Bouligand tangent cone $T(X,z)$ to X at z is defined as:*

$$T(X,z) = \{d : \exists \{z_n\} \subset X, \ z_n \to z, \ \exists \{\alpha_n\} \subset \Re, \ \alpha_n \to +\infty : \ \alpha_n(z_n - z) \to d\}.$$

Definition 2 *Let $X = \{x \in \Re^n : \ h_j(x) \leq 0, \ j = 1, \ldots, p\}$, h_j differentiable at $z \in X$. The linearizing cone to X at z is defined as:*

$$C(X,z) = \{d \in \Re^n : \ \nabla h_j(z)^T d \leq 0, \ j = 1, \ldots, p\}$$

$\Re^p_-$ and $\Re^p_+$ denote the non-positive and the non-negative orthant of $\Re^p$, respectively.

2 A fundamental separation theorem

In this section we will establish a separation theorem which allows us to develop a unifying approach for dealing with optimality conditions, constraint qualifications and theorem of alternatives.

Let W be a vector subspace of $\Re^p$; it is well known that if $W \cap int\Re^p_- = \emptyset$, then there exists a hyperplane which separates W and $int\Re^p_-$, i.e., there exists $\alpha \in \Re^p_+ \setminus \{0\}$ such that $\alpha^T w = 0$, for every $w \in W$. In general the vector α is not unique and it may happen that one (or more) component of α is zero whatever α be. In order to deep this aspect, we must investigate the intersection between W and the boundary of the cone $\Re^p_-$. With this aim we introduce the so-called conical extension $W^* = W + \Re^p_+$ which is a convex and closed cone.

The following lemma points out that the subspace W and its conical extension W^* have the same behaviour with respect to the intersection with $int\Re^p_-$.

Lemma 1 *Let W be a vector subspace of $\Re^p$. The following properties hold:*
i) $W \cap int\Re^p_- = \emptyset$ *if and only if* $W^* \cap int\Re^p_- = \emptyset$;
ii) a hyperplane Γ separates W and $\Re^p_-$ if and only if Γ separates W^* and $\Re^p_-$.

Proof. *i)* Since $W \subset W^*$, $W^* \cap int\Re^p_- = \emptyset$ implies $W \cap int\Re^p_- = \emptyset$. Assume now $W \cap int\Re^p_- = \emptyset$ and suppose the existence of $z \in W^* \cap int\Re^p_-$. We have $z = w + v$, $w \in W$, $v \in \Re^p_+$, so that $w = z - v \in int\Re^p_-$ since $z \in int\Re^p_-$; consequently $W \cap int\Re^p_- \neq \emptyset$, and this contradicts the assumption.
ii) Let Γ be a separating hyperplane of equation $\alpha^T z = 0$, $\alpha \geq 0$, so that $\alpha^T z = 0$, $\forall z \in W$. Let $w^* \in W^*$, i.e., $w^* = w + v$, $w \in W$, $v \in \Re^p_+$. We have $\alpha^T w^* = \alpha^T w + \alpha^T v = \alpha^T v \geq 0$, i.e., Γ separates W^* and $\Re^p_-$. The converse statement is obvious. ∎

From a geometrical point of view, when $W \cap int\Re^p_- = \emptyset$, the intersection between the conical extension of W and the non-positive orthant is a face $\mathcal{F}$ of $\Re^p_-$. The relevance of this property is due to the fact that it is possible to determine a set J of indices corresponding to multipliers which are zero in all separating hyperplanes and also to establish the existence of a separating hyperplane having positive multipliers associated with the indices which do not belong to J.

We recall that a face of $\Re^p_-$ is a set of the kind:

$$\mathcal{F} = \left\{ z = \sum_{j \in J} \gamma_j(-e^j), \ \gamma_j \geq 0 \right\}$$

where J is a proper subset of $\{1, ..., p\}$ and e^j, $j = 1, ..., p$, denotes the unit vector having the j-th component equal to one and all the others equal to zero. By convention $\mathcal{F} = \{0\}$ if $J = \emptyset$.

We have the following theorem.

Theorem 1 *Let W be a vector subspace of $\Re^p$ such that $W \cap int\Re^p_- = \emptyset$. Then, the following conditions hold:*

i) $W^* \cap \Re^p_-$ *is a face* $\mathcal{F} = \left\{ z = \sum_{j \in J} \gamma_j(-e^j), \ \gamma_j \geq 0 \right\}$, *where J is a proper subset of the set of indices* $\{1, ..., p\}$;
ii) if $J \neq \emptyset$, then, for every separating hyperplane of equation $\alpha^T z = 0$, $\alpha \geq 0$, we have $\alpha_j = 0$, $\forall j \in J$. Furthermore, there exists a separating hyperplane with $\alpha_i > 0$, $\forall i \notin J$;
iii) $J = \emptyset$ if and only if there exists a separating hyperplane with $\alpha_i > 0$, for every $i \in \{1, ..., p\}$.

Proof. *i)* If $W^* \cap \Re_-^p = \{0\}$, the thesis follows by convention, otherwise there exists $z \in W \cap \Re_-^p, z \neq 0$. Since $W \cap int\Re_-^p = \emptyset$, z is a boundary point of $\Re_-^p$ and thus there exists a proper subset of indices $J_z \subset \{1, ..., p\}$ such that $z = \sum_{j \in J_z} \gamma_j(-e^j), \gamma_j > 0$. Taking into account that $W^* = W + \Re_+^p$ is a convex cone, we have $\frac{1}{\gamma_k}\left(z + \sum_{j \in J_z, j \neq k} \gamma_j e^j\right) = -e^k \in W^*$, for every $k \in J_z$. By repeating this process for every element of $W \cap \Re_-^p$, we obtain a subset $J = \cup J_z$. Since $W \cap int\Re_-^p = \emptyset$, J is properly contained in $\{1, ..., p\}$. Consequently, the intersection $W^* \cap \Re_-^p$ is given by $\left\{z = \sum_{j \in J} \gamma_j(-e^j), \ \gamma_j \geq 0\right\}$, i.e., it is a face of $\Re_-^p$.

ii) Let $\alpha^T z = 0, \alpha \geq 0$, be the equation of a hyperplane which separates W and $\Re_-^p$; from ii) of Lemma 1, we have $\alpha^T w \geq 0, \forall w \in W^*$. Since $j \in J$ implies $-e^j \in W^*$, it results $\alpha^T(-e^j) = -\alpha_j \geq 0$, i.e., $\alpha_j \leq 0$, so that necessarily we have $\alpha_j = 0$ for every $j \in J$. Consider now the case $i \notin J$, so that $-e^i \notin W^*$. Since the convex cone W^* is the intersection of its supporting hyperplanes passing through the origin, there exists a hyperplane which separates W^* and $\Re_-^p$ and which does not contain $-e^i$. The equation of this hyperplane is of the kind $(\alpha^i)^T z = 0, \alpha^i \geq 0$, where necessarily $\alpha_i^i > 0$, and $\alpha_j^i = 0, \forall j \in J$. Let $\beta = \sum_{i \notin J} \alpha^i$. We have $\beta_j = 0, j \in J, \beta_i > 0, i \notin J$. Furthermore, since $(\alpha^i)^T z = 0, \forall z \in W$, we have $\beta^T z = 0, \forall z \in W$, and the proof is complete.

iii) $J = \emptyset$ if and only if $W^* \cap \Re_-^p = \{0\}$; this is equivalent to $-e^i \notin W^*$ for every $i \in \{1, ..., p\}$, i.e., to the existence of a separating hyperplane $\alpha^T z = 0$ with $\alpha_i > 0$, for every $i \in \{1, ..., p\}$. $\blacksquare$

Corollary 1 *Let W be a vector subspace of $\Re^p$ such that $W \cap int\Re_-^p = \emptyset$ and let $i \in \{1, ..., p\}$. The following statements are equivalent:*
i) $(-e^i) \notin W^*$;
ii) there exists a separating hyperplane of equation $\alpha^T z = 0, \alpha \geq 0$, such that $\alpha_i > 0$ and $\alpha^T w = 0, \forall w \in W$;
iii) there is no $w = (w_1, ..., w_i, ..., w_p) \in W$ such that $w_i < 0, w_j \leq 0, j \neq i$.

Proof. The thesis follows from ii) of Theorem 1 by noting that $i \in J$ if and only if $(-e^i) \in W^*$. On the other hand $(-e^i) \in W^*$ if and only if there exist $w \in W, v \in \Re_+^p$ such that $-e^i = w + v$; consequently, $w = (w_1, ..., w_i, ..., w_p)$ is such that $w_i < 0, w_j \leq 0, \ j \neq i$. $\blacksquare$

In order to clarify the properties stated in Theorem 1, consider the following simple example.

Example 1 *Let $W = \{t(-2, 0, -3, 0), \ t \in \Re\}$ be a subspace of $\Re^4$.*
We have $W \cap int\Re_-^4 = \emptyset$, and $W^* \cap \Re_-^4 = \{z = \gamma_1(-e^1) + \gamma_3(-e^3), \ \gamma_1, \gamma_3 \geq 0\}$. Conse-

quently, for every separating hyperplane of equation

$$\alpha_1 x_1 + \alpha_2 x_2 + \alpha_3 x_3 + \alpha_4 x_4 = 0, \ \alpha_j \geq 0, \ j = 1, ..., 4$$

necessarily we have $\alpha_1 = \alpha_3 = 0$. *Furthermore, it is possible to choose* $\alpha_2, \alpha_4 > 0$. *For instance* $x_2 + x_4 = 0$ *is a separating hyperplane; other separating hyperplanes are* $x_2 = 0$ *and* $x_4 = 0$.

When W is a Cartesian product, from Theorem 1 and from Corollary 1, we obtain the following fundamental result.

Theorem 2 *Let* W_1, W_2 *be vector subspaces of* $\Re^s$ *and* $\Re^m$, *respectively. Then, there exists a hyperplane of equation* $\alpha^T u + \beta^T p = 0$, *with* $\alpha \in int\Re^s_+$, $\beta \in \Re^m_+$, *which separates* $W_1 \times W_2$ *and* $\Re^s_- \backslash \{0\} \times \Re^m_-$ *if and only if*

$$(W_1 \times W_2) \cap (\Re^s_- \backslash \{0\} \times \Re^m_-) = \emptyset. \tag{1}$$

Proof. By setting $p = s + m$, the thesis follows from Theorem 1 taking into account that (1) is equivalent to state that $W_1 \times W_2$ does not contain elements of the kind (w^1, w^2) with $w^1_i < 0, i = 1, ..., s, \ w^1_j \leq 0, j \neq i, \ w^2 \leq 0.$ ■

3 Necessary optimality conditions and constraint qualifications

In this section we will emphasize the role of the fundamental separation Theorem 1 in deriving some constraint qualifications in vector optimization.

Let $F : \Re^n \rightarrow \Re^s$, $G : \Re^n \rightarrow \Re^m$ and consider the following vector optimization problem:

$$P : \begin{cases} \min \ F(x) \\ \quad x \in S \end{cases}$$

where $S = \{x \in \Re^n : \ G(x) \in \Re^m_-\}$.

Through this chapter, we will denote by $f_i, \ i = 1, ..., s, \ g_j \ j = 1, ..., m$, the components of the functions F and G respectively and by $F'(x_0)$, $G'(x_0)$ the Jacobian matrices of F and G evaluated at x_0.

We recall that a feasible point $x_0 \in S$ is said to be a local efficient point for P, if there exists a neighborhood I of x_0 such that

$$F(x) \notin F(x_0) + \Re^s_- \backslash \{0\}, \ \forall x \in S \cap I$$

As it is well known, the efficiency of x_0 is equivalent to the optimality of x_0 with respect to s scalar problems as it is stated in the following theorem.

Theorem 3 $x_0 \in S$ *is a local efficient point for* P *if and only if , for every* $i = 1, ..., s,$ x_0 *is a local minimum point for the scalar problem:*

$$P_i : \begin{cases} \min \quad f_i(x) \\ \quad x \in Q^i \end{cases}$$

where $Q^i = \{x \in \Re^n : f_k(x) \leq f_k(x_0), \ k = 1, ..., s, \ k \neq i, \ G(x) \in \Re^m_-\}.$

The following theorem expresses a necessary optimality condition which involves the Bouligand tangent cone.

Theorem 4 *If* x_0 *is a local efficient point for* P, *and* F, G *are differentiable at* x_0, *then we have*

$$F'(x_0)d \in \Re^s_+, \ \forall d \in \bigcap_{i=1}^{s} clcoT(Q^i, x_0). \tag{2}$$

Proof. Let $i \in \{1, ..., s\}$, and let $d \in T(Q^i, x_0)$. Then, there exist $\{x_n\} \subset Q^i, \ x_n \to x_0,$ $\{\alpha_n\} \subset \Re_+, \ \alpha_n \to +\infty,$ such that $\alpha_n(x_n - x_0) \to d.$
By Taylor's expansion, we have $f_i(x_n) - f_i(x_0) = \nabla f_i(x_0)^T (x_n - x_0) + o(\| x_n - x_0 \|),$
with $\dfrac{o(\| x_n - x_0 \|)}{\| x_n - x_0 \|} \to 0.$ Since $f_i(x_n) - f_i(x_0) \geq 0, \ \forall n,$ and $\lim_{n \to +\infty} \alpha_n(f_i(x_n) - f_i(x_0)) =$
$\lim_{n \to +\infty} \left[\nabla f_i(x_0)^T \alpha_n(x_n - x_0) + \alpha_n \| x_n - x_0 \| \frac{o(\|x_n - x_0\|)}{\|x_n - x_0\|} \right] = \nabla f_i(x_0)^T d,$ it results
$\nabla f_i(x_0)^T d \geq 0, \ \forall d \in T(Q^i, x_0).$
Let $w \in coT(Q^i, x_0).$ Then, there exist $d^h \in T(Q^i, x_0), \ h = 1, ..., k, \ \lambda_h \geq 0, \ h = 1, ..., k,$
$\sum_{h=1}^{k} \lambda_h = 1,$ such that $w = \sum_{h=1}^{k} \lambda_h d^h.$ Consequently, $\nabla f_i(x_0)^T w = \sum_{h=1}^{k} \lambda_h \nabla f_i(x_0)^T d^h \geq 0,$
and thus $\nabla f_i(x_0)^T w \geq 0, \ \forall w \in coT(Q^i, x_0).$
At last, let $z \in clcoT(Q^i, x_0).$ Then, there exists a sequence $\{w_n\} \subset coT(Q^i, x_0)$ such that $w_n \to z,$ with $\nabla f_i(x_0)^T w_n \geq 0.$ By the continuity of the scalar product, we have $\nabla f_i(x_0)^T z \geq 0$ for every $i \in \{1, ..., s\},$ so that (2) follows. ∎

The following theorem states the classical Fritz John optimality conditions.
In what follows, without loss of generality, we will assume that all the constraints are binding at a local efficient point for P.

Theorem 5 *If* x_0 *is a local efficient point for problem* P, *and* F, G *are differentiable at* x_0, *then*

$$\exists (\lambda, \mu) \in \Re^s_+ \times \Re^m_+, \ (\lambda, \mu) \neq (0, 0) : \lambda^T F'(x_0) + \mu^T G'(x_0) = 0. \tag{3}$$

Proof. Consider the subspaces $W_1 = \{F'(x_0)d, \ d \in \Re^n\},$ $W_2 = \{G'(x_0)d, \ d \in \Re^n\}.$ It is easy to prove that the efficiency of x_0 implies $(W_1 \times W_2) \cap (int\Re^s_- \times int\Re^m_-) = \emptyset,$ so that the thesis follows applying Lemma 1. ∎

In general, it may happen that some multipliers related to the objectives are zero and consequently, it is not possible to deduce the behaviour at x_0 of the objective functions associated with such zero multipliers. Therefore, it is important to find necessary and/or sufficient conditions which guarantee $\lambda \in int\Re_+^s$ in (3).

When $\lambda \in int\Re_+^s$, Fritz John conditions are referred to as Karush-Kuhn-Tucker conditions and any condition which ensures $\lambda \in int\Re_+^s$ is called constraint qualification.

Now, as a direct consequence of Theorem 2, we are able to state a necessary and sufficient condition for the validity of Fritz John conditions (3) with the positivity of all the multipliers associated with the objectives. Such a condition can be interpreted as a very general constraint qualification.

The following theorem holds.

Theorem 6 *Let x_0 be a local efficient point for problem P, and suppose that F, G are differentiable at x_0. Then, there exist $(\lambda, \mu) \in int\Re_+^s \times \Re_+^m$, such that*

$$\lambda^T F'(x_0) + \mu^T G'(x_0) = 0$$

if and only if

$$(W_1 \times W_2) \cap (\Re_-^s \setminus \{0\} \times \Re_-^m) = \emptyset \tag{4}$$

where $W_1 = \{F'(x_0)d, \ d \in \Re^n\}$, $W_2 = \{G'(x_0)d, \ d \in \Re^n\}$.

Remark 1 *We would like to point out that, in general, the Karush-Kuhn-Tucker conditions are obtained by applying suitable theorems of the alternative which allow to prove that $\lambda \in int\Re_+^s$ in (3), if and only if the following system is impossible:*

$$\begin{cases} \nabla f_i(x_0)^T d \leq 0 & i = 1, ..., s \\ \nabla f_i(x_0)^T d < 0 & \text{at least one } i \\ \nabla g_j(x_0)^T d \leq 0 & j = 1, ..., m \end{cases}$$

As we will see in Section 5, there is a strict connection between the fundamental separation Theorem 1 and the theorems of the alternative. In our geometric approach, we are able to specify the sign of the multipliers by looking for the edges of the non-positive orthant which do not belong to the conic extension $W^* = (W_1 \times W_2) + \Re_+^{s+m}$.

In vector optimization, unlike in the scalar case, any constraint qualifications, suggested in the literature, involves both the constraint set and the objective functions. This is due to the fact that, in a multiobjective problem, when the linearizing cone coincides with the closure of the convex hull of the Bouligand tangent cone to the feasible set, we do not necessarily have $\lambda \in int\Re_+^s$ in (3). This is pointed out in Example 2.

Example 2 *Consider problem P, where $f(x_1, x_2) = (x_1 - x_2, \ x_2^3)$ and the feasible set is $S = \{(x_1, x_2) \in \Re^2 : x_1 \geq 0\}$. It is easy to verify that $x_0 = (0, 0) \in S$ is an efficient point and that Fritz John conditions become:*

$$\lambda_1 \begin{pmatrix} 1 \\ -1 \end{pmatrix} + \lambda_2 \begin{pmatrix} 0 \\ 0 \end{pmatrix} + \mu \begin{pmatrix} 1 \\ 0 \end{pmatrix} = \begin{pmatrix} 0 \\ 0 \end{pmatrix}, \ (\lambda_1, \lambda_2, \mu) \in \Re_+^3 \setminus \{(0, 0, 0)\} \tag{5}$$

Consequently, (5) is satisfied by $\lambda_1 = 0$, $\lambda_2 > 0$, $\mu = 0$, so that $\lambda = (0, \lambda_2) \notin int\Re^2_+$. Nevertheless, $clcoT(S, x_0) = T(S, x_0) = C(S, x_0) = S$.

In [14], Maeda generalizes the Guignard constraint qualification to the vector case. In Theorem 7 and in Example 3, we will point out that this relevant condition is sufficient but not necessary for the positivity of all multipliers associated with the objectives in (3). Let

$$Q = \{x \in \Re^n : G(x) \in \Re^m_-, F(x) - F(x_0) \in \Re^s_- \setminus \{0\}\}.$$

When $s = 1$, we set $Q = S$.

Theorem 7 *Let x_0 be a local efficient point for problem P, where F, G are differeentiable at x_0, and assume that*

$$C(Q, x_0) = \bigcap_{i=1}^{s} clcoT(Q^i, x_0) \tag{6}$$

Then, condition (4) holds, i.e.,

$$(W_1 \times W_2) \cap (\Re^s_- \setminus \{0\} \times \Re^m_-) = \emptyset \tag{7}$$

Proof. From Theorem 4, we have $F'(x_0)d \in \Re^s_+$, $\forall d \in \bigcap_{i=1}^{s} clcoT(Q^i, x_0)$. On the other hand, $F'(x_0)d \in \Re^s_-$, $\forall d \in C(Q, x_0)$, so that the constraint qualification (6) implies $F'(x_0)d = 0$, $\forall d \in C(Q, x_0)$. Consequently, (4) holds. ∎

Example 3 *Consider problem P, where $f(x_1, x_2) = \left(x_1^3 - x_2, x_1 x_2 - 3x_2\right)$ and $S = \{(x_1, x_2) \in \Re^2 : g_1(x_1, x_2) = x_1^2 - x_2 \leq 0, g_2(x_1, x_2) = x_2 \leq 0\}$.*
It is easy to verify that $S = \{(0, 0)\}$, so that $x_0 = (0, 0)$ is efficient for P. By simple calculation, we have $C(Q, x_0) = \{(d_1, 0) : d_1 \in \Re\}$ and $\bigcap_{i=1}^{2} clcoT(Q^i, x_0) = \{(0, 0)\}$, so that the constraint qualification (6) does not hold.
On the other hand, condition (4) holds since $W_1 \cap \Re^2_+ \setminus \{0\} = \emptyset$ and therefore Karush-Khun-Tucker conditions are verified (for instance, we can choose $\lambda_1 = \lambda_2 = \mu_1 = 1$ and $\mu_2 = 5$).

As we have previously pointed out, in vector optimization, any constraint qualification must involve both the objective functions and the constraints, so we need to consider the sets Q^i and Q.

With this respect, it is important to observe that, when x_0 is a local efficient point for problem P, we have $intC(Q, x_0) = \emptyset$. It follows that every constraint qualification for a scalar problem, involving $intC \neq \emptyset$ cannot be extended to the vector case. Nevertheless, following the same line of the scalar case, it can be proven (see [14]) that the following conditions are constraint qualifications:

Abadie constraint qualification: $C(Q, x_0) = T(Q, x_0)$

Generalized Abadie constraint qualification: $C(Q, x_0) = \bigcap_{i=1}^{s} T(Q^i, x_0)$

Cottle-type constraint qualification: $int C(Q^i, x_0) \neq \emptyset, \; \forall i$.

Remark 2 *In [14], the following constraint qualification is suggested:*

$$rank F'(x_0) = s$$

and the system

$$\begin{cases} F'(x_0)d = 0 \\ G'(x_0)d \in int \Re^m \end{cases}$$

has a solution $d \in \Re^n$.

We point out (see also [8]) that such a condition cannot be considered as a constraint qualification since it is not verified when x_0 is a local efficient point. In fact, the efficiency of x_0 implies the validity of the Fritz John conditions and the existence of d satisfying (3) implies $\lambda^T F'(x_0)d + \mu^T G'(x_0)d = \mu^T G'(x_0)d = 0$, so that necessarily, $\mu = 0$. It follows that $\lambda^T F'(x_0) = 0$ with $\lambda \in \Re_+^s \setminus \{0\}$ and this is a contradiction since $rank F'(x_0) = s$, i.e., the rows of $F'(x_0)$ are linearly independent.

Unlike the scalar case, the linear independence of the gradients of the constraints does not guarantee $\lambda \in int \Re_+^s$ in (3) as it is shown in Example 4; nevertheless, adding to this linear independence, the indipendence of $s - 1$ gradients of the objective functions, we obtain a new constraint qualification, stated in Theorem 8.

Example 4 *Consider problem P, where $f(x_1, x_2) = (x_1^2 + x_2^2 - 2x_1 - 2x_2, x_2)$, and $S = \{(x_1, x_2) : g_1(x_1, x_2) = x_1 + x_2 \leq 0\}$.*
It can be verified that $x_0 = (0, 0)$ is efficient for P and that conditions (3) are satisfied if and only if $\lambda = (\lambda_1 > 0, \lambda_2 = 0)$, $\mu = 2\lambda_1$, so that Karush-Kuhn-Tucker conditions do not hold.

Theorem 8 *Let x_0 be a local efficient point for problem P, assume that f_i, $i = 1, ..., s$, g_j, $j = 1, ..., m$ are differentiable at x_0, and let $F_{-i} = (f_1, ..., f_{i-1}, f_{i+1}, ..., f_s)$.*
If $rank \begin{bmatrix} F'_{-i}(x_0) \\ G'(x_0) \end{bmatrix} = s - 1 + m$, $i = 1, ..., s$, then (4) holds.

Proof. Assume that (4) is not verified. Then, there exists $d \in \Re^n$ such that $F'(x_0)d \in \Re_+^s \setminus \{0\}$, $G'(x_0)d \in \Re_-^m$. Let $i \in \{1, ..., s\}$ be such that $\nabla f_i(x_0)^T d < 0$; the efficiency of x_0 implies that x_0 is a local minimum point for the scalar problem P_i. By assumption, the gradients of the constraints of P_i, evaluated at x_0, are linearly independent, so that, for problem P_i, the Karush-Kuhn-Tucker conditions hold, i.e., there exist $\lambda \in \Re_+^{s-1}$, $\mu \in \Re_+^m$ such that

$$\nabla f_i(x_0) + \lambda^T F_i'(x_0) + \mu^T G'(x_0) = 0.$$

On the other hand, $\nabla f_i(x_0)^T d < 0$, $\lambda^T F_i'(x_0)d \in \Re_-^{s-1}$, $\mu^T G'(x_0)d \in \Re_-^m$ imply

$$(\nabla f_i(x_0) + \lambda^T F_i'(x_0) + \mu^T G'(x_0))^T d < 0$$

and this is a contradiction. ∎

4　Generalized convexity and constraint qualifications

In the scalar case several constraint qualifications involve generalized convexity assumptions. An extension of these results to the vector case can be found in [8, 12, 14, 19] where generalized convexity is required for any components of the objective functions and/or of the constraint functions.

In order to obtain new constraint qualifications, we will consider classes of vector generalized convex functions which are more general than the componentwise generalized convex ones.

As is known, there are different ways in extending generalized convexity to the vector case and there are many definitions proposed in the literature (see for instance [5, 6, 13, 17]). We will address to those classes which will allow us to obtain constraint qualifications.

Let K be a convex subset of an open set $X \subseteq \Re^n$. Let $H : X \to \Re^p$, and assume that H is differentiable at $x_0 \in K$.

We will consider the following classes:

- H is said to be $int\Re_-^p$-pseudoconvex at $x_0 \in K$ if

$$x \in K, \ \ H(x) \in H(x_0) + int\Re_-^p \Rightarrow H'(x_0)(x - x_0) \in int\Re_-^p$$

- H is is said to be pseudolinear at $x_0 \in K$ if the following two statements hold:

$$x \in K, \ \ H(x) \in H(x_0) + \Re_-^p \setminus \{0\} \Leftrightarrow H'(x_0)(x - x_0) \in \Re_-^p \setminus \{0\}$$

$$x \in K, \ \ H(x) \in H(x_0) + \Re_+^p \setminus \{0\} \Leftrightarrow H'(x_0)(x - x_0) \in \Re_+^p \setminus \{0\}$$

Furthermore, we will introduce the following new class:

- H is said to be reverse pseudoconcave at $x_0 \in K$ if

$$x \in K, \ \ H'(x_0)(x - x_0) \in \Re_-^p \ \Rightarrow \ H(x) \in H(x_0) + \Re_-^p$$

When $p = 1$, $int\Re_-^p$-pseudoconvexity (pseudolinearity) reduces to pseudoconvexity[2](pseudolinearity), while reverse pseudoconcavity reduces to pseudoconcavity.

It can be easily proved that the class of the componentwise pseudoconvex (pseudoconcave, pseudolinear) functions is contained in the class of $int\Re_-^p$-pseudoconvex (reverse pseudoconcave, pseudolinear) functions. More precisely we have the following results.

Theorem 9 *Let $H = (h_1, ..., h_p)$.*
i) If h_i, $i = 1, ..., p$, are pseudoconvex at x_0, then H is $int\Re_-^p$-pseudoconvex at x_0.
ii) If h_i, $i = 1, ..., p$, is pseudoconcave at x_0, then H is reverse pseudoconcave at x_0.
iii) If h_i, $i = 1, ..., p$, is pseudolinear at x_0, then H is pseudolinear at x_0.

The following examples show that the inclusion relationships, stated in Theorem 9 are strict.

Example 5 *Consider the function $H(x) = (x_1, x_1 + x_2^3, x_2)$, $x = (x_1, x_2)^T$, and the point $x_0 = (0, 0)$. H is $int\Re_-^3$-pseudoconvex at x_0. In fact, we have $H(x) \in H(x_0) + \Re_-^3$ if and only if $x_1 < 0$, $x_2 < 0$, so that $H'(x_0)(x - x_0) = (x_1, x_1, x_2)^T \in int\Re_-^3$.*
On the other hand, $h_2(x) = x_1 + x_2^3$ is not pseudoconvex at x_0, since $h_2(1, -2) = -7 < h_2(0, 0) = 0$ but $\nabla h_2(x_0)(1, -2) = 1 > 0$.

Example 6 *Consider the function $H(x) = (x_1, x_1^3, x_2)$, $x = (x_1, x_2)^T$, and the point $x_0 = (0, 0)$. H is reverse pseudoconcave at x_0. In fact, we have $H'(x_0)(x - x_0) = (x_1, 0, x_2)^T \in \Re_-^3$ if and only if $x_1 \leq 0$, $x_2 \leq 0$. It follows that $H(x) \in H(x_0) + \Re_-^3$. On the other hand, $h_2(x) = x_1^3$ is not pseudoconcave at x_0.*

Example 7 *Consider the function $H(x) = (x^3 + x, x^3)$, and the point $x_0 = 0$. It is easy to verify that H is pseudolinear at x_0, but $h_2(x) = x^3$ is not pseudolinear at x_0.*

As a direct consequence of Theorem 9, we have the following corollary.

Corollary 2 *Let $H = (h_1, ..., h_p)$.*
i) If h_i, $i = 1, ..., p$, are convex or linear at x_0, then H is $int\Re_-^p$-pseudoconvex at x_0.
ii) If h_i, $i = 1, ..., p$, are concave or linear at x_0, then H is reverse pseudoconcave at x_0.

Consider now the vector function H_{-i} obtained from $H = (h_1, ..., h_p)$ by deleting the i-th component, i.e., $H_{-i} = (h_1, ..., h_{i-1}, h_{i+1}, ..., h_p)$.
Observe that the $int\Re_-^p$-pseudoconvexity of H does not imply necessarily the pseudoconvexity of H_{-i}, $i = 1, ..., p$, as it is shown in the following example.

---

[2]Let $h : S \to \Re$, where S is an open convex subset of $\Re^n$, and assume that h is differentiable at $x_0 \in S$.
• h is pseudoconvex at x_0 if $x \in S$, $h(x) < h(x_0)$ implies $\nabla h(x_0)^T (x - x_0) < 0$.
• h is pseudoconcave at x_0 if and only if $-h$ is pseudoconvex.
• h is pseudolinear at x_0 if it is both pseudoconvex at x_0 and pseudoconcave at x_0.

Example 8 *Consider the function H, defined in Example 5. H is $int\Re^3_-$-pseudoconvex at $x_0 = (0,0)$, but the function $H_{-1}(x) = (x_1 + x_2^3, x_2)$ is not $int\Re^2_-$-pseudoconvex at x_0, since we have $H_{-1}(1,-2) = (-7,-2) \in int\Re^2_-$, and $H'_{-1}(x_0)(1,-2)^T = (1,-2)^T \notin int\Re^2_-$.*

Now we will establish some constraint qualifications involving the classes of generalized convex functions introduced before.

Let x_0 be a local efficient point for problem P, where F and G are differentiable at x_0, and $G(x_0) = 0$. Consider the following conditions:

- **Condition 1.** The objective function F and the constraint function G are reverse pseudoconcave at x_0.

- **Condition 2.** Let I be a proper subset of $\{1,...,m\}$ and set $I^* = \{1,...,m\} \setminus I$. F and G_I are reverse pseudoconcave at x_0 and there exists $d \in \Re^n$, $d \neq 0$ such that $G'_I(x_0)d \in \Re^{|I|}_-$, $G'_{I^*}(x_0)d \in int\Re^{|I^*|}_-$, $F'(x_0)d \in \Re^s_-$, where $|I|$, $|I^*|$ denote the cardinality of the sets I, I^*, respectively.

- **Condition 3.** F is reverse pseudoconcave at x_0 and there exists $d \in \Re^n$, $d \neq 0$ such that $G'(x_0)d \in int\Re^m_-$, $F'(x_0)d \in \Re^s_-$.

- **Condition 4.** F_{-i} is $int\Re^{s-1}_-$-pseudoconvex at x_0, $i = 1,...,s$, and G is $int\Re^m_-$-pseudoconvex at x_0. Moreover, for every $i = 1,...,s$, there exists $\bar{x}^{(i)}$ such that $F_{-i}(\bar{x}^{(i)}) \in F_{-i}(x_0) + int\Re^{s-1}_-$, $G(\bar{x}^{(i)}) \in G(x_0) + int\Re^m_-$.

- **Condition 5.** F and G are pseudolinear at x_0 and

$$G'(y)(x-y) = 0 \Leftrightarrow G(x) = G(y) \tag{8}$$

The following theorem holds.

Theorem 10 *Every condition 1-5 is a constraint qualification.*

Proof. Referring to Theorem 6, we will give the proof by assuming that (4) does not hold or, equivalently, by assuming the existence of a vector $\bar{d} \neq 0$, such that

$$F'(x_0)\bar{d} \in \Re^s_- \setminus \{0\}, \; G'(x_0)\bar{d} \in \Re^m_- \tag{9}$$

Condition 1 The reverse pseudoconcavity assumption on the constraint function G implies that $\bar{d}$ is a feasible direction starting from x_0, while the reverse pseudoconcavity of F implies $F(x_0 + t\bar{d}) \in F(x_0) + \Re^s_-$ for every $t > 0$ such that $x_0 + t\bar{d} \in S$. Furthermore, there exists i such that $\nabla f_i(x_0)\bar{d} < 0$, so that function f_i decreases along the direction $\bar{d}$ and the efficiency of x_0 is contradicted.

Condition 2 Let $d^* = \bar{d} + d$; we have $G'_I(x_0)d^* \in \Re^{|I|}_-$, and $G'_{I^*}(x_0)d^* \in int\Re^{|I^*|}_-$, so that d^* is a feasible direction starting from x_0. Since $F'(x_0)d^* \in \Re^s_- \setminus \{0\}$, we contradict the efficiency of x_0.

Condition 3 This is a particular case of Condition 2.

Condition 4 Let j be such that $\nabla f_j(x_0)\bar{d} < 0$ and consider F_{-j}.

From assumption, there exists $\bar{x}^{(j)}$ such that $F_{-j}(\bar{x}^{(j)}) \in F_{-j}(x_0) + int\Re_-^{s-1}$, and $G(\bar{x}^{(j)}) \in G(x_0) + int\Re_-^m$.

Since F_{-j} is $int\Re_-^{s-1}$-pseudoconvex at x_0 and G is $int\Re_-^m$-pseudoconvex at x_0, setting $d = \bar{x}^{(j)} - x_0$, we have $F'_{-j}(x_0)d \in int\Re_-^{s-1}$, and $G'(x_0)d \in int\Re_-^m$.

Let $\hat{d} = \bar{d} + \frac{1}{n}d$. We have $G'(x_0)\hat{d} = G'(x_0)\bar{d} + \frac{1}{n}G'(x_0)d \in int\Re_-^m$, $\forall n$, so that $\hat{d}$ is a feasible direction.

On the other hand, $F'_{-j}(x_0)\hat{d} = F'_{-j}(x_0)\bar{d} + \frac{1}{n}F'_{-j}(x_0)d \in int\Re_-^{s-1}$, $\forall n$, and, for n large enough, $\nabla f_j(x_0)\hat{d} = \nabla f_j(x_0)\bar{d} + \frac{1}{n}\nabla f_j(x_0)d < 0$, so that we have $F'(x_0)\hat{d} \in int\Re_-^s$ and this contradicts the efficiency of x_0.

Condition 5 Let $d = x - x_0$, $x \in \Re^n$; we have $G'(x_0)(x - x_0) \in \Re_-^m$ if and only if $G'(x_0)(x - x_0) = 0$ or $G'(x_0)(x - x_0) \in \Re_-^m \setminus \{0\}$.

If $G'(x_0)(x - x_0) = 0$, then (8) implies $G(x) = G(x_0)$ so that $x \in S$.

If $G'(x_0)(x - x_0) \in \Re_-^m \setminus \{0\}$, the pseudolinearity of G implies $G(x) \in \Re_-^m \setminus \{0\}$ and, once again, $x \in S$.

On the other hand, the pseudolinearity of F implies $F(x) \in F(x_0) + \Re_-^s \setminus \{0\}$ and this contradicts the efficiency of x_0. ∎

Remark 3 *In the scalar case, i.e., $s = 1$, Condition 1 reduces to the weak-reverse constraint qualification, Condition 2 and 3 coincide with weak Arrow-Hurwicz-Uzawa constraint qualification, while Condition 4 is equal to weak Slater constraint qualification (see for all [15]).*

5   Theorems of the alternative

In this section, we will show how Theorem 1 and Corollary 1 can be utilized for deriving, in a very simple way, some known theorems of the alternative.

Let A, B, C be given matrices of order $m \times n, p \times n, q \times n$, respectively.

Set:

- $W_1 = \{w_1 = Ax, \ x \in \Re^n\}$, $W_2 = \{w_2 = Bx, \ x \in \Re^n\}$, $W_3 = \{w_3 = Cx, \ x \in \Re^n\}$,

- $W = \{w = (w_1, w_2, w_3)^T, \ w_1 \in W_1, \ w_2 \in W_2, \ w_3 \in W_3\} \subset \Re^{m+p+q}$,

- $I_1 = \{1, ..., m\}$, $I_2 = \{m + 1, ..., m + p\}$, $I_3 = \{m + p + 1, ..., m + p + q\}$,

- $e^i, \ i \in \{1, ..., m + p + q\}$, the unit vector of $\Re^{m+p+q}$.

Since many theorems of the alternative have been appeared in the literature, we have chosen the ones which allow us to point out how the separation Theorem 1 works.

Farkas' lemma

Farkas' lemma is used extensively in deriving optimality conditions of linear and nonlinear programming problems. The lemma can be stated as follows.

Let A be an $m \times n$ matrix and $b \in \Re^m$ a non null vector.

Then, exactly one of the following two systems has a solution:

System 1 $Ax \in \Re^m_-$, $b^T x > 0$

System 2 $\alpha^T A = b^T$, $\alpha \in \Re^m_+$.

Proof. Consider the set W with $B = -b^T$, and $W_3 = \emptyset$.

If System 2 has a solution, then we have $\alpha^T Ax = b^T x$, so that $Ax \in \Re^m_-$ implies $b^T x \leq 0$. Consequently, System 1 is impossible.

Assume now that System 1 has no solutions. Then, $(w_1, w_2) \notin W$, with $w_1 \in \Re^m_-$, $w_2 < 0$, and thus $W \cap int\Re^m_- = \emptyset$. Note that $-e^{m+1} \notin W^* = W + \Re^{m+1}_+$, otherwise there exists $w \in W$ such that $w \in -e^{m+1} - \Re^{m+1}_-$, and this contradicts the impossibilty of System 1. From Theorem 1 it follows the existence of $\alpha_1 \in \Re^m_+$, $\beta > 0$ such that $\alpha_1^T A - \beta^T I = 0$. Consequently, System 2 is verified with $\alpha = \dfrac{\alpha_1}{\beta}$. ∎

Remark 4 *Note that nothing can be said about the positivity of the components of the multiplier vector α, since an element $w = (w_1, 0) \in \Re^{m+1}_-$ can or cannot belong to W.*

Gale's Theorem 1

Let A be an $m \times n$ matrix.

Then, exactly one of the following two systems has a solution:

System 1 $Ax \in \Re^m_- \setminus \{0\}$, $x \in \Re^n_+$

System 2 $\alpha^T A \in \Re^n_+$, $\alpha \in int\Re^m_+$.

Proof. Consider the set W with $B = -I$ and $W_3 = \emptyset$.

If System 2 has a solution, $x \in \Re^n_+$ implies $\alpha^T Ax > 0$, so that, taking into account the assumption, we have $\alpha \in int\Re^m_+$, $Ax \notin \Re^m_-$, i.e., System 1 is impossible.

Assume now that System 1 has no solutions. Then, $(w_1, w_2) \notin W$, with $w_1 \in \Re^m_-$, $w_2 \in -\Re^n_+$, and thus $W \cap int\Re^{m+n}_- = \emptyset$. Note that for every $i \in I_1$, $-e^i \notin W^*$, otherwise there exists $w \in W$ such that $w \in -e^i - \Re^{m+n}_-$, and this contradicts the impossibility of System 1. From Theorem 1 it follows the existence of $\alpha \in int\Re^m_+, \beta \in \Re^n_+$ such that $\alpha^T A - \beta^T I = 0$ or, equivalently, $\alpha^T A = \beta^T \in \Re^n_+$. ∎

Gale's Theorem 2

Let A be an $m \times n$ matrix, and $b \in \Re^n \setminus \{0\}$.

Then, exactly one of the following two systems has a solution:

System 1 $Ax \in \Re^m_-$, $b^T x < 0$, $x \in \Re^n_+$

System 2 $\alpha^T A + b^T \in \Re^n_+ \setminus \{0\}$, $\alpha \in \Re^m_+$.

Proof. Consider the set W with $B = b^T, C = -I$.

If System 2 has a solution, $x \in \Re^n_+$, and $b^T x < 0$ imply $(\alpha^T A + b^T)x \geq 0$, i.e., $\alpha^T Ax \geq -b^T x > 0$. On the other hand, $Ax \in \Re^m_-$ and $\alpha \in \Re^m_+$ imply $\alpha^T Ax \leq 0$. Consequently, System 1 is impossible.

Assume now that System 1 has no solutions. Then, $(w_1, w_2, w_3) \notin W$, with $w_1 \in \Re_-^m$, $w_2 < 0$, $w_3 \in \Re_-^n$, and thus $W \cap int\Re_-^{m+n} = \emptyset$. Note that $-e^i \notin W^*, i \in I_2$, otherwise there exists $w \in W$ such that $w \in -e^i - \Re_-^{m+1+n}$, and this contradicts the impossibility of System 1. From Theorem 1 it follows the existence of $\alpha_1 \in \Re_+^m, \beta > 0, \gamma \in \Re_+^n$ such that $\alpha_1^T A + \beta^T b - \gamma^T = 0$ or, equivalently, $(\alpha_1^T A + \beta^T) \in \Re_+^n$. ∎

Slater's Theorem

Let A, B, C be given matrices of order $m \times n, p \times n, q \times n$, respectively.

Then, exactly one of the following two systems has a solution:

System 1 $Ax \in int\Re_-^m$, $Bx \in \Re_-^p \setminus \{0\}$, $Cx \in \Re_-^q$

System 2 $\alpha^T A + \beta^T B + \gamma^T C = 0$ with

$\alpha \in \Re_+^m \setminus \{0\}, \; \beta \in \Re_+^p, \gamma \in \Re_+^q$ or $\alpha \in \Re_+^m, \; \beta \in int\Re_+^p, \gamma \in \Re_+^q$.

Proof. Consider the set W.

If System 2 has a solution, we have $\alpha^T Ax + \beta^T Bx + \gamma^T Cx = 0$, for every $x \in \Re^n$, while $Ax \in int\Re_-^m$, $\alpha \in \Re_+^m \setminus \{0\}$ or $Bx \in \Re_-^p \setminus \{0\}$, $\beta \in \Re_+^p$, imply $\alpha^T Ax + \beta^T Bx + \gamma^T Cx < 0$. Consequently, System 1 is impossible.

Assume now that System 1 has no solutions. Then, $(w_1, w_2, w_3) \notin W$, with $w_1 \in int\Re_-^m$, $w_2 \in \Re_-^p \setminus \{0\}$, $w_3 \in \Re_-^q$, and thus $W \cap int\Re_-^{m+p+q} = \emptyset$. Note that $(w_1, w_2, 0) \notin W^*$, with $w_1 \in int\Re_-^m$, $w_2 = -\bar{e}^j$, where $\bar{e}^j$ denotes the unit vector of $\Re^p$, otherwise there exists $w \in W$ such that $w \in (w_1, w_2, 0) - \Re_-^{m+p+q}$, and this contradicts the impossibility of System 1. It follows that the face $\mathcal{F} = \{z = \sum_{k \in I_1 \cup I_2} \gamma_k(-e^k), \; \gamma_k \geq 0\}$ is not contained in W^*, so that at least one edge of the face does not belong to W^* or, equivalently, either $-e^j \notin W^*$, $j = m+1, ..., m+p$ or $-e^i \notin W^*$ for at least one $i \in I_1$. Consequently, from Theorem 1, there exist α, β, γ satisfying System 2. ∎

6   Conclusions

In this chapter, by means of a new separation theorem, we have presented a necessary and sufficient condition for having the positivity of the multipliers associated with the objective functions in Fritz John necessary optimality conditions. According to the strict relationship between constraint qualifications and theorems of the alternative, we have underlined that our separation theorem is a useful tool in deriving theorems of the alternative. At last we have deepened the role of vector generalized convex functions in establishing constraint qualifications.

We would like to thank Professor Alberto Cambini for his helpful and stimulating comments and suggestions which have greately improved the obtained results.

References

[1] M.S. Bazaraa, H.D. Sheraly and C.M. Shetty, Non linear programming, second edition, John Wiley & Sons, Inc., New York, 1993.

[2] A. Cambini and L. Martein, Generalized concavity in multiobjective programming, in Generalized convexity, Generalized monotonicity, Proceedings of the "Fifth International Symposium on Generalized Convexity" edited by Crouzeix J.P., Martnez-Legaz J.E., and Volle M., Nonconvex Optim. Appl., 27, Kluwer Academic Publishers, New York, (1998)453-467.

[3] A. Cambini and L. Martein, Generalized Convexity and Optimality Conditions in Scalar and Vector Optimization, in Handbook of Generalized convexity and Generalized Monotonicity edited by Hadjisavvas N., Komlòsi S., and Schaible S., Springer, Verlag Berlin, 2005.

[4] A. Cambini and L. Martein, Generalized Convexity and Optimization, Theory and Applications, Lecture Notes in Economics and Mathematical Systems, Vol. 616, Springer, 2009.

[5] R. Cambini, Some new classes of generalized concave vector-valued functions, Optimization 36(1996)11-24.

[6] R. Cambini L. Martein, First and Second Order Characterizations of a Class of Pseudo-concave Vector Functions, in Generalized Convexity and Generalized Monotonicity, edited by Hadjisavvas N., Martinez-Legaz J.E. and Penot J.P., Lecture Notes in Economics and Mathematical Systems, Vol. 502, Springer, Berlin (2001) 144-158.

[7] S. Chandra , J. Dutta and C.S. Lalitha, Regularity Conditions and Optimality in Vector Optimization, Numerical Functional Analysis and Optimization 25(2004)479-501.

[8] G. Giorgi, B. Jiménez and V. Novo, On constraint qualifications in directionally differentiable multiobjective optimization problems, RAIRO Operations Research 38(2004)255-274.

[9] C. Fulga and V. Preda, On optimality conditions for multiobjective optimization problems in topological vector space, Journal of Mathematical Analysis and Applications 334(2007)123-131.

[10] B. Jiménez and V. Novo, Alternative theorems and necessary optimality conditions for directionally differentiable multiobjective optimization programs, Journal of Convex Analysis 9(2002)97-116.

[11] X.F. Li, Constraint Qualifications in Nonsmooth Multiobjective Optimization, Journal of Optimization Theory and Applications 106(2000)373-398.

[12] X.F. Li and J.Z. Zhang, Stronger Kuhn-Tucker Type Conditions in Nonsmooth Multiobjective Optimization: Locally Lipschitz Case, Journal of Optimization Theory and Applications 127(2005)367-388.

[13] D.T. Luc, Theory of vector optimization, Springer-Verlag, Germany, 1989.

[14] T. Maeda, Constraint Qualifications in Multiobjective Optimization Problems: Differentiable Case, Journal of Optimization Theory and Applications 80(1994)483-499.

[15] O.L. Mangasarian, Nonlinear programming, SIAM Philadelphia, 1994.

[16] L. Martein, Some Results on Regularity in Vector Optimization, Optimization 20(1989)787-798.

[17] S.K. Mishra and G. Giorgi, Invexity and Optimization, Springer, 2008.

[18] M.D. Nguyen and D.V. Luu, On necessary conditions for efficiency in directionally differentiable optimization problems, Cahier de la MSE, Centre dEconomie de la Sorbonne UMR 8174, 2006.

[19] V. Preda and I. Chitescu, On Constraint Qualifications in Multiobjective Optimization Problems: Semidifferentiable Case, Journal of Optimization Theory and Applications 100(1999)417-433.

Optimality Conditions in Vector Optimization, 2010, 35-60

CHAPTER 3

Second order optimality conditions in vector optimization problems.

M. Hachimi*   B. Aghezzaf†

Abstract

We are interested in proving optimality conditions for optimization problems. By means of different second-order tangent sets, various second-order necessary optimality conditions are obtained for both scalar and vector optimization problems where the feasible region is given as a set. We present also second-order sufficient optimality conditions so that there is only a very small gap with the necessary optimality conditions. As an application we establish second-order optimality conditions of Fritz John type, Kuhn-Tucker type 1, and Kuhn-Tucker type 2 for a problem with both inequality and equality constraints and a twice differentiable functions. At the end, a very general second-order necessary conditions for efficiency with respect to cones is present and it is applied to smooth and nonsmooth data.

Keywords: Multiobjective programming, second-order tangent sets, constraint qualifications, second-order optimality conditions, efficient solutions.

1 Introduction

The investigation of the optimality conditions is one of the most attractive topics of optimization theory. As far as we know, the classical paper [29] of Kuhn and Tucker has been the first paper to state an optimality condition for a differentiable problem by using the first-order derivatives of vector functions. Several extensions of this condition have been achieved for larger classes of problems, including problems with nonsmooth data or set-valued data. For many optimization problems, the characterization of optimal solutions with help of second order conditions was always of a great interest of refine first optimality conditions. The need of second order informations also appears in numerical algorithms.

Examination of most theoretical results on the first-order necessary conditions, in vector optimization, reveals that Lin's fundamental theorem [30] is usually a source for establishing the results (see e.g. [39]). Naturally, an interesting question will arise: Who

---

*Professor, Universié Ibn Zohr, Faculté des Sciences Juridiques Economiques et Sociales, B.P. 8658 Hay Dakhla - Agadir 80000 Morocco. e-mail: hachimi@lycos.com

†Professor, Département de Mathématiques et d'Informatique, Faculté des Sciences, Université Hassan II Aïn chock, B.P.5366 - Maârif Casablanca Morocco. e-mail: b.aghezzaf@fsac.ac.ma

first-order optimality conditions can be extended to second-order ones? In recent years, there has been an increasing interest in the generalization of the second-order optimality conditions in multiobjective optimization problems and different approaches have been suggested (see e.g. [2, 7, 9, 10, 11, 12, 14, 22, 23, 41]). In [3], we have generalized the Lin's fundamental theorem via a second-order tangent set which can be considered an extension of the tangent cone and obtained second-order necessary optimality conditions for efficiency. However, it is possible to build other second-order tangent sets by extending the tangent cone. Cambini *et al.* ([13]) and Penot ([37]) introduce a new second-order tangent set called the asymptotic second-order cone, which allows them to reduce the gap between necessary and sufficient conditions to an acceptable extent. Taking into account the importance of the asymptotic second-order cone, we have refined in [26] necessary conditions obtained recently (see e.g. [2, 3, 9, 10, 11, 14]) and present new second-order optimality conditions for both scalar and vector optimization problems.

The outline of this chapter is as follows. In Section 2, the notations are introduced and some preliminary results are given. In Section 3, we first derive the second-order necessary optimality conditions for a point to be a local (weak) efficient point of an arbitrary subset of the criteria space. We derive also second-order sufficient optimality conditions so that the gap with the necessary ones is reduced to an acceptable extent. Second, we establish second-order necessary optimality conditions for a point to be a local (weak) efficient solution of a vector optimization problem with an arbitrary set (of the decision space) and a twice continuously differentiable objective function. We explain also the gap between vector and scalar optimization; we deduce second-order necessary optimality conditions for scalar optimization problems. Using the relationship between the vector optimization problem and its corresponding individual scalar problems, we derive other second-order necessary optimality conditions for a point to be a local (weak) efficient solution of the vector optimization problem. In Section 4, the feasible set is defined by inequality and equality constraints. Firstly, without any constraint qualification, we obtain second-order necessary conditions of Fritz John. Secondly, with constraint qualification, we obtain second-order necessary conditions of Kuhn-Tucker type 1. Finally, under a regularity condition, we derive second-order necessary conditions of Kuhn-Tucker type 2. In Section 5, we prove second-order sufficient conditions of Fritz John, Kuhn-Tucker type 1, and Kuhn-Tucker type 2 under weak invexity assumptions. Finally, in Section 6, we prove very general second-order necessary optimality conditions for a point to be a local (weak) efficient point, with respect to an ordering convex pointed cone, of an arbitrary subset of the criteria space. Then, we apply such conditions to obtain second-order necessary optimality conditions for multiobjective problems with an arbitrary feasible set and an objective function for which the gradient map is a smooth or nonsmooth function.

2   Preliminaries

In this section, we shall introduce notations and definitions which are used throughout the paper. For α, $\beta \in \mathbb{R}$, by $\alpha \leqslant \beta$, we mean α is inferior or equal to β. Let $\mathbb{R}^n$ be the n-dimensional Euclidean space. For x, $y \in \mathbb{R}^n$, by $x \leqq y$, we mean $x_i \leqslant y_i$ for all i; by $x \leq y$, we mean $x \leqq y$ and $x \neq y$; and by $x < y$, we mean $x_i < y_i$ for all i. We denote the inner product of x and y by $xy = x^t y = x_1 y_1 + \cdots + x_n y_n$; here the superscript denotes the transpose of x. We denote by $\mathbb{R}^n_-$ the set of vectors $x \in \mathbb{R}^n$ which satisfy $x_i \leqslant 0$ for all i.

For each $S \subset \mathbb{R}^n$, the sets $\operatorname{int} S$, $\overline{S}$, $\operatorname{Fr} S$, $\operatorname{cone} S$ and $\operatorname{co} S$ denote the interior, the closure, the boundary, the generated cone and the convex hull of S, respectively. We denote by $B(\bar{x}, \varepsilon)$ the open ball centered at $\bar{x}$ and of radius ε.

For any two vectors $x = (x_1, x_2)$ and $y = (y_1, y_2)$ in $\mathbb{R}^2$, $x \leq_{\text{lex}} y$ means that $x_1 < y_1$ holds or $x_1 = y_1$ and $x_2 \leqslant y_2$. Similarly, $x <_{\text{lex}} y$ means that $x_1 < y_1$ holds or $x_1 = y_1$ and $x_2 < y_2$; here the subscript lex is an abbreviation of lexicographic order.

Now, for any twice continuously differentiable vector function $g : \mathbb{R}^n \longrightarrow \mathbb{R}^m$ and for any vectors $x, y \in \mathbb{R}^n$, we denote by $\nabla g(x)$ and $\nabla^2 g(x)(y, y)$ the $m \times n$ Jacobian matrix and the m-dimensional vector whose ith component is $y^t \nabla^2 g_i(x) y$.

In this paper, we study a multiobjective optimization problem of the form

$$(\text{MOP}) \qquad \text{minimize } f(x) \quad \text{subject to} \quad x \in S$$

where $f : \mathbb{R}^n \longrightarrow \mathbb{R}^\ell$ is a function and S is a subset of the decision space $\mathbb{R}^n$. We do not yet fix the form of the constraint functions forming S, but refer to S in general. We will refer to $\mathbb{R}^n$ as the decision space and to $\mathbb{R}^\ell$ as the criteria space.

Due to the conflicting nature of the objectives, a minimizer solution that simultaneously minimizes all the objectives is usually not obtainable. We shall give the following concepts of solutions to problem (MOP). For other notions and their connections see [31] and [43].

Definition 1 *A point $\bar{x} \in S$ is called to be a local efficient [resp. local weak efficient] solution to problem (MOP) if there exists a neighbourhood N of $\bar{x}$ such that no $x \in S \cap N$ satisfies $f(x) \leq f(\bar{x})$ [resp. $f(x) < f(\bar{x})$].*

If the above definition holds with $N = \mathbb{R}^n$, the point $\bar{x}$ is called efficient [resp. weak efficient] solution to problem (MOP). We also define efficiency in the criteria space.

Definition 2 *A point $\bar{y} \in Z \subset \mathbb{R}^\ell$ is called to be a local efficient [resp. local weak efficient] point of Z if there exists a neighbourhood V of $\bar{y}$ such that no $y \in Z \cap V$ satisfies $y \leq \bar{y}$ [resp. $y < \bar{y}$].*

Also, if the above definition holds with $N = \mathbb{R}^\ell$, the point $\bar{y}$ is called efficient [resp. weak efficient] point of Z.

Now, we focus on certain approximation sets to the feasible region. The following definition of tangent cone is used by Penot in [36]

Definition 3 *Let S be a nonempty subset of $\mathbb{R}^n$. The tangent cone to S at $\bar{x} \in \overline{S}$ is the set defined by*

$$T(S, \bar{x}) = \left\{ d \in \mathbb{R}^n \mid \exists\, t_n \longrightarrow 0^+,\ \exists\, d_n \longrightarrow d \ \text{ such that } \ x_n = \bar{x} + t_n d_n \in S \right\} \quad (1)$$

For more properties of this type of approximation, see [27].

Next, we shall define two second-order approximations of the feasible region S at $\bar{x} \in \overline{S}$ which can be considered as two different extensions of the tangent cone $T(S, \bar{x})$. These approximations are very useful for the formulation of second-order optimality conditions

Definition 4 *Let S be a nonempty subset of $\mathbb{R}^n$. The second-order tangent set to S at $\bar{x} \in \overline{S}$ is the set defined by*

$$T^2(S, \bar{x}) = \Big\{ (d, z) \in \mathbb{R}^n \times \mathbb{R}^n \mid \exists\, t_n \longrightarrow 0^+,\ \exists\, z_n \longrightarrow z \ \text{ such that }$$

$$x_n = \bar{x} + t_n d + (1/2) t_n^2 z_n \in S \Big\} \quad (2)$$

Definition 5 *Let S be a nonempty subset of $\mathbb{R}^n$. The asymptotic second-order tangent cone to S at $\bar{x} \in \overline{S}$ is the set defined by*

$$T''(S, \bar{x}) = \Big\{ (d, z) \in \mathbb{R}^n \times \mathbb{R}^n \mid \exists\, t_n \longrightarrow 0^+,\ \exists\, r_n \longrightarrow 0^+,\ \exists\, z_n \longrightarrow z$$

$$\text{such that } \ t_n / r_n \longrightarrow 0, \quad x_n = \bar{x} + t_n d + (1/2) r_n t_n z_n \in S \Big\} \quad (3)$$

By definition, it follows immediately that the asymptotic second-order tangent cone is in fact a cone.

Other second-order approximations of the feasible region S at $\bar{x} \in \overline{S}$ can be inspired by a notion given in the work of Cambini *et al.* in [13]. Let k be any real number, we define the second-order tangent set of index k to S at $\bar{x} \in \overline{S}$ by:

$$T''_k(S, \bar{x}) = \Big\{ (d, z) \in \mathbb{R}^n \times \mathbb{R}^n \mid \exists\, t_n \longrightarrow 0^+,\ \exists\, r_n \longrightarrow 0^+,\ \exists\, z_n \longrightarrow z$$

$$\text{such that } \ t_n / r_n \longrightarrow k, \quad x_n = \bar{x} + t_n d + (1/2) r_n t_n z_n \in S \Big\} \quad (4)$$

The d-sections of $T''_k(S, \bar{x})$ will be denoted by

$$T''_k(S, \bar{x})(d) = \{ z \in \mathbb{R}^n \mid (d, z) \in T''_k(S, \bar{x}) \}$$

Remark 1 *It is easy to show that $T''_0(S, \bar{x}) = T''(S, \bar{x})$ and $T''_1(S, \bar{x}) = T^2(S, \bar{x})$. Moreover, for every $k \geqslant 0$, we observe that $T''_k(S, \bar{x})(0) = T(S, \bar{x})$ and if $d \notin T(S, \bar{x})$ then $T''_k(S, \bar{x})(d) = \varnothing$.*

The following proposition shows the relationship between the second-order tangent set of index $k > 0$ and the classical second-order tangent set.

Proposition 1 *Let S be a subset of X and let $\bar{x} \in \overline{S}$. Then, for all $k > 0$, $T''_k(S, \bar{x}) = (1/k)\, T^2(S, \bar{x})$. In particular, $T''_1(S, \bar{x}) = T^2(S, \bar{x})$.*

    Proof.   It follows immediately from the definition of $T''_k(S, \bar{x})$. $\square$

3　Geometric optimality conditions

In this section, we establish various second-order optimality conditions with the aid of the second-order tangent set and the asymptotic second-order tangent cone, both in the general case and in the differentiable case. First, we derive a second-order necessary optimality conditions in the criteria space which allow us to deduce the rest.

Theorem 1 *If $\bar{y}$ is a local (weak) efficient point of $Z \subset \mathbb{R}^\ell$, then for every $k \in \{0, 1\}$ the following holds*

$$T_k''(Z, \bar{y}) \cap \Omega = \varnothing, \tag{5}$$

where　$\Omega = \{(d, z) \in \mathbb{R}^\ell \times \mathbb{R}^\ell \mid (d_i, z_i)^t <_{\text{lex}} (0, 0)^t, \forall i \}$

Proof.　Suppose to the contrary that (5) does not hold for some $k \in \{0, 1\}$, that is there exist $(t_n, r_n) \longrightarrow (0^+, 0^+)$ and $z_n \longrightarrow z$ such that $t_n/r_n \longrightarrow k$ and

$$y_n = \bar{y} + t_n d + (1/2) r_n t_n z_n \in Z. \tag{6}$$

If $d_i < 0$, one has $r_n t_n (z_n)_i = o(t_n d_i)$ so that $(y_n)_i - (\bar{y})_i < 0$ when n is sufficiently large. If $d_i = 0$, one has $(z_n)_i < 0$. Hence for n large $(y_n)_i - (\bar{y})_i = (1/2) r_n t_n (z_n)_i < 0$. Consequently, $y_n < \bar{y}$ for n sufficiently large which contradicts the hypothesis of the theorem.　□

Remark 2 *As shown in the proof of Theorem 1, the condition (5) holds for every real number k. However, nothing essential is lost when k is restricted by condition $k \in \{0, 1\}$. In fact, if $k < 0$ then $T_k''(Z, \bar{y})$ is empty and (5) holds trivially; and if $k > 0$, we have the following equivalence*

$$T_1''(Z, \bar{y}) \cap \Omega = \varnothing \Longleftrightarrow (1/k) \, T_1''(Z, \bar{y}) \cap (1/k) \, \Omega = \varnothing \Longleftrightarrow T_k''(Z, \bar{y}) \cap \Omega = \varnothing$$

because Ω is a cone, and $(1/k) \, T_1''(Z, \bar{y}) = T_k''(Z, \bar{y})$ by Proposition 1.

Corollary 1 *If $\bar{y}$ is a local (weak) efficient point of $Z \subset \mathbb{R}^\ell$, then for every direction $d \in T(Z, \bar{y}) \cap \text{Fr}(\mathbb{R}_-^\ell)$ the following holds*

$$T_0''(Z, \bar{y})(d) \cap \Omega(d) = \varnothing, \tag{7a}$$
$$T_1''(Z, \bar{y})(d) \cap \Omega(d) = \varnothing, \tag{7b}$$

where　$\Omega(d) = \{z \in \mathbb{R}^\ell \mid (d, z) \in \Omega \}$

Proof.　It follows immediately from Theorem 1.　□

It may happen that either (7a) or (7b) fails to detect nonefficient points, as the following examples shows

Example 1 *We consider the following set Z used by Cambini et al in [13].*

$$Z = \{ (y_1, y_2) \in \mathbb{R}^2 \mid y_2 = -y_1 \sqrt{|y_1|} \}, \quad where \quad \bar{y} = (0,0) \quad and \quad d = (-1,0)$$

Let $z = (z_1, z_2)$, it can be seen that,

$$\Omega(d) = \{z \in \mathbb{R}^2 \mid z_2 < 0\}, \quad T_1''(Z,\bar{y})(d) = \varnothing, \quad T_0''(Z,\bar{y})(d) = \{z \in \mathbb{R}^2 \mid z_2 \geqslant 0\},$$

Hence, condition (7a) is satisfied, but (7b) is false. So $\bar{y}$ is not efficient point of Z.

Example 2 *We consider the following set Z.*

$$Z = \{ (y_1, y_2) \in \mathbb{R}^2 \mid y_2 = -y_1^2 \}, \quad where \quad \bar{y} = (0,0) \quad and \quad d = (-1,0)$$

Let $z = (z_1, z_2)$, it can be seen that,

$$T_1''(Z,\bar{y})(d) = \{z \in \mathbb{R}^2 \mid z_2 = -2\}, \quad T_0''(Z,\bar{y})(d) = \{z \in \mathbb{R}^2 \mid z_2 = 0\},$$

Here, condition (7a) is false, but (7b) is satisfied. So $\bar{y}$ is not efficient point of Z.

Although the second-order set $T_1''(Z,\bar{y})(d)$ may be empty, as the example 1 shows, the following result asserts that the union of $T_0''(Z,\bar{y})(d)$ and $T_1''(Z,\bar{y})(d)$ is always nonempty.

Proposition 2 *Let $d \in T(Z,\bar{y})$, where Z is an arbitrary subset of $\mathbb{R}^\ell$ and $\bar{y} \in \overline{Z}$. Then either $T_0''(Z,\bar{y})(d)$ or $T_1''(Z,\bar{y})(d)$ is nonempty.*

    Proof.   The proof is similar to that of Proposition 2.1 of Penot [37].   □

    Lin ([30]) has established sufficient conditions for local efficiency. We will extend Lin's Theorem 5.2 by means of second-order tangent sets.

Theorem 2 *A sufficient condition for a point $\bar{y} \in Z \subset \mathbb{R}^\ell$ to be a local efficient point of Z is the condition:*

$$T(Z,\bar{y}) \cap (\mathbb{R}_-^\ell) \subseteq \mathrm{Fr}\,(\mathbb{R}_-^\ell) \tag{8}$$

and for any non vanishing d in $T(Z,\bar{y}) \cap \mathrm{Fr}\,(\mathbb{R}_-^\ell)$, if it exists, we have

$$T_0''(Z,\bar{y})(d) \cap d^\perp \cap \overline{\Omega(d)} = \{0\} \tag{9a}$$

$$T_1''(Z,\bar{y})(d) \cap d^\perp \cap \overline{\Omega(d)} = \varnothing. \tag{9b}$$

where $d^\perp$ denotes the orthogonal subspace to d.

    Proof.   First let us assume that $T(Z,\bar{y}) \cap (\mathbb{R}_-^\ell) = \{0\}$. Then $\bar{y}$ is a local efficient point of Z by Theorem 5.2 in [30]. Now let us assume that $T(Z,\bar{y}) \cap (\mathbb{R}_-^\ell)$ is strictly larger than $\{0\}$ and that condition (9) holds, but $\bar{y}$ was not a local efficient point of Z. By definition 2, there would exist a sequence $\{y_n\} \subset Z$ such that

$$y_n \leq \bar{y} \quad and \quad y_n \longrightarrow \bar{y} \tag{10}$$

we can assume that $y_n = \bar{y} + t_n \, d_n$, where $\|d_n\| = 1$, $d_n \longrightarrow d$ and $t_n = \|y_n - \bar{y}\|$. Thus $d \in T(Z, \bar{y})$ with $\|d\| = 1$. It is obvious that $d \in \mathbb{R}_-^\ell$. Taking into account condition (8) we conclude that $d \in T(Z, \bar{y}) \cap \mathrm{Fr}\,(\mathbb{R}_-^\ell)$ and $d \neq 0$.

Setting $z_n = 2(d_n - d)/t_n$. So, we have the following

$$d_n = d + (1/2)t_n z_n \qquad \text{and} \qquad y_n = \bar{y} + t_n d + (1/2)t_n^2 z_n \tag{11}$$

We consider the case when $\{z_n\}$ is bounded and alternatively when it is not bounded. First, we assume that $\{z_n\}$ is bounded. Then, taking a subsequence if necessary, there exist z such that $z_n \longrightarrow z$. So, from (11), we get $z \in T_1''(Z, \bar{y})(d) \cap T(C, d)$ where C is defined by $C = \{v \in \mathbb{R}^\ell \mid \|v\| = 1\}$. Since $T(C, d) = d^\perp$, we have $z \in T_1''(Z, \bar{y})(d) \cap d^\perp$. On the other hand, it is easy to show that $z \in \overline{\Omega(d)}$ which contradicts condition (9b).

If we now assume that $\{z_n\}$ is not bounded. Then, taking a subsequence if necessary, we can suppose that $\|z_n\| \longrightarrow +\infty$ and that there exists $z' \neq 0$ such that $z_n' = \|z_n\|^{-1} z_n \longrightarrow z'$. From (11) we get

$$d_n = d + (1/2)r_n z_n' \qquad \text{and} \qquad y_n = \bar{y} + t_n d + (1/2)r_n t_n z_n' \tag{12}$$

where $r_n = \|z_n\| t_n = 2\|d_n - d\| \longrightarrow 0^+$, and $t_n/r_n \longrightarrow 0$. This implies that $z' \in T_0''(Z, \bar{y})(d) \cap d^\perp$. It is also easy to show that $z' \in \overline{\Omega(d)}$ which again contradicts condition (9a). $\qquad\qquad\square$

The following example shows a case in which Lin's Theorem 5.2 is not applicable. However Theorem 2 can be applied.

Example 3 *We consider the set Z defined by*

$$Z = \{\, (y_1, y_2) \in \mathbb{R}^2 \mid y_2 = y_1 \sqrt{|y_1|}\,\}, \quad \text{with} \ \ \bar{y} = (0, 0),$$

It can be seen that,

$$T(Z, \bar{y}) = \{(d_1, d_2) \in \mathbb{R}^2 \mid d_2 = 0\},$$

and for each $d \in T(Z, \bar{y})$ and $d \neq 0$ we have

$$T_1''(Z, \bar{y})(d) = \varnothing, \quad T_0''(Z, \bar{y})(d) = \{(z_1, z_2) \in \mathbb{R}^2 \mid z_2 \geqslant 0\}.$$

It is easy to show that condition (8) holds. Calculations show also that conditions (9) hold which implies that $\bar{y}$ is an efficient point of Z.

In what follows we assume that the objective function f is twice continuously differentiable. Now, we are in position to give second-order necessary optimality conditions in the decision space.

Theorem 3 *If $\bar{x}$ is a local (weak) efficient solution for problem* (MOP), *then the system*

$$(\nabla f_l(\bar{x})d, \nabla f_l(\bar{x})z + k\nabla^2 f_l(\bar{x})(d, d)) <_{\mathrm{lex}} (0, 0), \quad \forall\, l,$$

has no solution (d, z) in $T_k''(S, \bar{x})$ for every $k \in \{0, 1\}$.

Proof. According to definitions 1 and 2, it is easy to see that if $\bar{x}$ is a local (weak) efficient solution to problem (MOP), then there exists a neighborhood N of $\bar{x}$ such that $f(\bar{x})$ is (weak) efficient point of $f(S \cap N)$. On the other hand, for any fixed k, let (d, z) be any element of $T_k''(S, \bar{x})$. Then, taking into account $T_k''(S, \bar{x}) = T_k''(S \cap N, \bar{x})$, there exist $(t_n, r_n) \longrightarrow (0^+, 0^+)$ and $z_n \longrightarrow z$ such that $t_n/r_n \longrightarrow k$ and

$$x_n = \bar{x} + t_n d + (1/2)r_n t_n z_n \in S \cap N \tag{13}$$

By Taylor's expansion, there exist $\varepsilon_n \longrightarrow 0$ such that

$$
\begin{aligned}
f(x_n) &= f(\bar{x}) + t_n \nabla f(\bar{x})(d + (1/2)r_n z_n) \\
&\quad + (1/2)t_n^2 \nabla^2 f(\bar{x})(d + (1/2)r_n z_n, d + (1/2)r_n z_n) + t_n^2 \varepsilon_n
\end{aligned}
$$

Thus, we have

$$
\begin{aligned}
f(x_n) &- f(\bar{x}) + t_n \nabla f(\bar{x})d + (1/2)t_n r_n \Big[\nabla f(\bar{x})(z_n) \\
&\quad + k_n \nabla^2 f(\bar{x})\left(d + (1/2)r_n z_n, d + (1/2)r_n z_n\right) + 2k_n \varepsilon_n \Big]
\end{aligned}
$$

with $k_n \longrightarrow k$. Since f is twice continuously differentiable, then

$$\nabla f(\bar{x})(z_n) + k_n \nabla^2 f(\bar{x})\left(d + (1/2)r_n z_n, d + (1/2)r_n z_n\right) \xrightarrow[n \to +\infty]{} \nabla f(\bar{x})z + k\nabla^2 f(\bar{x})(d, d)$$

which implies that

$$(\nabla f(\bar{x})d, \nabla f(\bar{x})z + k\nabla^2 f(\bar{x})(d, d)) \in T_k''(f(S \cap N), f(\bar{x})).$$

Since $f(\bar{x})$ is a (weak) efficient point of $f(S \cap N)$, Theorem 1 leads to

$$T_k''(f(S \cap N), f(\bar{x})) \cap \Omega = \varnothing.$$

Hence

$$(\nabla f(\bar{x})d, \nabla f(\bar{x})z + k\nabla^2 f(\bar{x})(d, d)) \notin \Omega.$$

This complete the proof. $\qquad \square$

Remark 3 *It is easy to show that the set Λ defined by*

$$\Lambda = \{(d, z) \in \mathbb{R}^{2n} \mid (\nabla f_l(\bar{x})d, \nabla f_l(\bar{x})z + k\nabla^2 f_l(\bar{x})(d, d)) <_{\text{lex}} (0, 0), \quad \forall\, l\}$$

is a cone. Hence, for each $k > 0$, the following holds

$$\Lambda \cap T_1''(S, \bar{x}) = \varnothing \iff \Lambda \cap (1/k)\, T_1''(S, \bar{x}) = \varnothing \iff \Lambda \cap T_k''(S, \bar{x}) = \varnothing.$$

Definition 6 *The set of the critical directions for problem (MOP) at $\bar{x} \in \overline{S}$ is the set defined by*

$$\mathbb{K} = \Big\{ d \in \mathbb{R}^n \mid d \in T(S, \bar{x}),\ \nabla f(\bar{x})d \leq 0,\ \nabla f_i(\bar{x})d = 0, \quad \text{at least one } i \Big\} \tag{14}$$

For any feasible point $\bar{x} \in S$ and $d \in \mathbb{R}^n$, we set

$$I_f = \{1, \ldots, \ell\}, \qquad I_f(\bar{x}, d) = \{l \in I_f \mid \nabla f_l(\bar{x})d = 0\} \tag{15}$$

Corollary 2 *If $\bar{x}$ is a local (weak) efficient solution to problem* (MOP), *then for each critical direction $d \in \mathbb{K}$ the system*

$$\nabla f_l(\bar{x})z + k\nabla^2 f_l(\bar{x})(d, d) < 0, \quad l \in I_f(\bar{x}, d), \tag{16}$$

has no solution z in $T_k''(S, \bar{x})(d)$ for every $k \in \{0, 1\}$.

Proof. It follows immediately from Theorem 3 $\qquad\qquad\qquad\qquad\qquad$ $\square$

Now, we consider the scalar optimization problems and we extend Corollary 2 to such problems. The following lemma is useful (See [5]) and will make things easy. It's proof is obvious and is omitted here.

Lemma 1 *Let Γ and T be sets in $\mathbb{R}^n$. Suppose that Γ is a open half space. Then,*

$$T \cap \Gamma = \varnothing \Longrightarrow \overline{\mathrm{co}}\,(T) \cap \Gamma = \varnothing,$$

where $\overline{\mathrm{co}}\,(T)$ is the closure of the convex hull of T.

Now, we can state second-order optimality conditions for scalar optimization problems.

Theorem 4 *Let $\ell = 1$. If $\bar{x}$ is a (local) minimizer solution to problem* (MOP), *then for each critical direction $d \in \mathbb{K}$ the system*

$$\nabla f(\bar{x})z + k\nabla^2 f(\bar{x})(d, d) < 0$$

has no solution z in $\overline{\mathrm{co}}\,[T_k''(S, \bar{x})(d)]$ for every $k \in \{0, 1\}$.

Proof. For $k \in \{0, 1\}$, let

$$\Gamma = \{z \in \mathbb{R}^n \mid \nabla f(\bar{x})z + k\nabla^2 f(\bar{x})(d, d) < 0\}.$$

The set Γ may be either empty or $\mathbb{R}^n$ or open half space. In each cases Lemma 1 holds. Consequently, the result follows from Corollary 2 and Lemma 1. $\qquad\qquad$ $\square$

Now, let us return to vector optimization problems. Before deriving a variant of Corollary 2, we needs the following notations and lemma.

For each $i = 1, 2, \ldots, \ell$, we shall define the nonempty sets Q^i and Q by

$$Q^i = \{x \in \mathbb{R}^n \mid x \in S, \quad f_l(x) \leqslant f_l(\bar{x}), \quad l = 1, 2, \ldots, \ell \quad \text{and} \quad l \neq i\} \tag{17}$$

$$Q = \{x \in \mathbb{R}^n \mid x \in S, \quad f_l(x) \leqslant f_l(\bar{x}), \quad l = 1, 2, \ldots, \ell\} \tag{18}$$

and the ith objective constraint problem (P^i) by

$$(\mathrm{P}^i) \qquad \text{minimize } f_i(x) \quad \text{subject to} \quad x \in Q^i \tag{19}$$

The following lemma gives the relationship between the multiobjective problem (MOP) and the corresponding individual scalar problems

Lemma 2 *A point $\bar{x}$ is (local) efficient solution to problem* (MOP) *if and only if it is a (local) minimizer to problem* (P^i) *for every* $i = 1, \ldots, \ell$

    Proof. It follows immediately from proof of Theorem 3.4.11 in [38]. $\square$

Theorem 5 *If $\bar{x}$ is (local) efficient solution to problem* (MOP), *then for each critical direction $d \in \mathbb{K}$ the system*

$$\nabla f_l(\bar{x})z + k\nabla^2 f_l(\bar{x})(d, d) < 0, \quad \text{for at least one } l \in I_f(\bar{x}, d) \tag{20}$$

has no solution z in $\bigcap_{i=1}^{\ell} \overline{\mathrm{co}}\left[T_k''(Q^i, \bar{x})(d)\right]$ for every $k \in \{0, 1\}$.

    Proof. It follows immediately from Theorem 4 and Lemma 2 $\square$

    The following theorem shows that we can obtain sufficient optimality conditions by replacing strict inequalities in (16) by inequalities.

Theorem 6 *A sufficient condition for a point $\bar{x} \in S$ to be a local efficient solution to problem* (MOP) *is the condition:*

$$T(S, \bar{x}) \cap \{d \in \mathbb{R}^n \mid \nabla f(\bar{x})d \leqq 0\} \subseteq \mathbb{K} \tag{21}$$

and for any critical direction $d \in \mathbb{K}$ with $d \neq 0$, if it exists, the following two systems

$$\begin{aligned} \nabla f_l(\bar{x})z \leqslant 0, \quad l \in I_f(\bar{x}, d), \\ z \in T_0''(S, \bar{x})(d) \cap d^\perp, \quad z \neq 0 \end{aligned}$$

and

$$\begin{aligned} \nabla f_l(\bar{x})z + \nabla^2 f_l(\bar{x})(d, d) \leqslant 0, \quad l \in I_f(\bar{x}, d), \\ z \in T_1''(S, \bar{x})(d) \cap d^\perp \end{aligned}$$

have no solution $z \in \mathbb{R}^n$.

    Proof. The proof of this theorem makes use the arguments of Theorem 2. $\square$

4   John and Kuhn-Tucker necessary conditions

In what follows we consider the feasible region S defined by inequality and equality constraints. Let $g : \mathbb{R}^n \longrightarrow \mathbb{R}^p$ and $h : \mathbb{R}^n \longrightarrow \mathbb{R}^q$ be twice continuously differentiable functions and let $I_g = \{1, \ldots, p\}$ and $I_h = \{1, \ldots, q\}$ be the corresponding index sets. We will denote the feasible region by

$$S = \{x \in \mathbb{R}^n \mid g(x) \leqq 0, \quad h(x) = 0\}$$

For any feasible point $\bar{x} \in S$ and $d \in \mathbb{R}^n$, we set

$$I_g(\bar{x}) = \{i \in I_g \mid g_i(\bar{x}) = 0\}, \qquad I_g(\bar{x}, d) = \{i \in I_g(\bar{x}) \mid \nabla g_i(\bar{x})d = 0\}$$
$$I_f = \{1, \ldots, \ell\}, \qquad I_f(\bar{x}, d) = \{l \in I_f \mid \nabla f_l(\bar{x})d = 0\}$$

Definition 7 *The kth second-order linearizing set to S at $\bar{x}$ is*

$$
\begin{aligned}
L_k''(S, \bar{x}) \;=\; & \{(d, z) \in \mathbb{R}^{2n} \;| \\
& (\nabla g_i(\bar{x})d, \nabla g_i(\bar{x})z + k\nabla^2 g_i(\bar{x})(d,d)) \leqq_{\text{lex}} (0,0), \quad i \in I_g(\bar{x}), \\
& (\nabla h_j(\bar{x})d, \nabla h_j(\bar{x})z + k\nabla^2 h_j(\bar{x})(d,d)) = (0,0), \quad j \in I_h \}
\end{aligned}
$$

Definition 8 *The kth weak second-order linearizing set to S at $\bar{x}$ is*

$$
\begin{aligned}
WL_k''(S, \bar{x}) \;=\; & \{(d, z) \in \mathbb{R}^{2n} \;| \\
& (\nabla g_i(\bar{x})d, \nabla g_i(\bar{x})z + k\nabla^2 g_i(\bar{x})(d,d)) <_{\text{lex}} (0,0), \quad i \in I_g(\bar{x}), \\
& (\nabla h_j(\bar{x})d, \nabla h_j(\bar{x})z + k\nabla^2 h_j(\bar{x})(d,d)) = (0,0), \quad j \in I_h \}
\end{aligned}
$$

In order to derive second-order necessary optimality conditions for problem (MOP), we shall proof the following lemmas, which shows the relationship between the second-order tangent sets and the second-order linearizing sets

Lemma 3 *Let $\bar{x} \in S$ be any feasible solution to problem* (MOP). *Then,*

$$
T_k''(S, \bar{x}) \subseteq L_k''(S, \bar{x}), \quad k \in \{0, 1\} \tag{22}
$$

If moreover $\{\nabla h_j(\bar{x}), \quad j \in I_h\}$ are linearly independent, then

$$
WL_k''(S, \bar{x}) \subseteq T_k''(S, \bar{x}), \quad k \in \{0, 1\} \tag{23}
$$

Proof. It is similar to the proof of Lemma 4.1 in [9]. See also [25]   □
It follows immediately from (22) that, for every direction $d \in \mathbb{R}^n$

$$
T_k''(S, \bar{x})(d) \subseteq L_k''(S, \bar{x})(d), \quad k \in \{0, 1\} \tag{24}
$$

where $L_k''(S, \bar{x})(d) = \{z \in \mathbb{R}^n \mid (d, z) \in L_k''(S, \bar{x})\}$. Since $L_k''(S, \bar{x})(d)$ is a closed convex set for each $d \in \mathbb{R}^n$, (24) leads to

$$
\overline{\text{co}}\,[T_k''(S, \bar{x})(d)] \subseteq L_k''(S, \bar{x})(d), \quad k \in \{0, 1\} \tag{25}
$$

Lemma 4 *Let $\bar{x} \in S$ be any feasible solution to problem* (MOP). *Then,*

$$
\bigcap_{i=1}^{\ell} \overline{\text{co}}\,[T_k''(Q^i, \bar{x})(d)] \subseteq L_k''(Q, \bar{x})(d), \quad k \in \{0, 1\} \tag{26}
$$

Proof. Taking into account the fact that $\bigcap_{i=1}^{\ell} L_k''(Q^i, \bar{x})(d) = L_k''(Q, \bar{x})(d)$, the proof follows from condition (25).   □

4.1   John type necessary optimality conditions

Now we reduce the geometric necessary optimality conditions of Corollary 2 in terms of the gradients and Hessians of the objective and the constraint functions.

Theorem 7 *Let* $k \in \{0,1\}$. *Suppose that* $\{\nabla h_j(\bar{x}), \quad j \in I_h\}$ *are linearly independent. If* $\bar{x}$ *is a local (weak) efficient solution to problem* (MOP), *then for each critical direction* $d \in \mathbb{K}$ *the system*

$$\nabla f_l(\bar{x})z + k\nabla^2 f_l(\bar{x})(d,d)) < 0, \quad l \in I_f(\bar{x}, d) \tag{27a}$$
$$\nabla g_i(\bar{x})z + k\nabla^2 g_i(\bar{x})(d,d)) < 0, \quad i \in I_g(\bar{x}, d) \tag{27b}$$
$$\nabla h_j(\bar{x})z + k\nabla^2 h_j(\bar{x})(d,d)) = 0, \quad j \in I_h \tag{27c}$$

has no solution z *in* $\mathbb{R}^n$.

Proof. It follows from Corollary 2 and condition (23) of Lemma 3 □
In order to obtain multiplier rules, the following lemma will be useful.

Lemma 5 *Let* A, B *and* C *be given real matrices, with* $A \neq 0$. *Let* a, b *and* c *be real vectors. If the system*

$$(\Upsilon) \qquad Ax + a < 0, \qquad Bx + b \leqq 0, \qquad Cx + c = 0 \tag{28}$$

has no solution x *in* $\mathbb{R}^n$, *then exactly one of the two following systems* (I) *and* (II) *has a solution:*

$$(\text{I}) \qquad \lambda^t A + \mu^t B + \nu^t C = 0$$
$$\lambda^t a + \mu^t b + \nu^t c \geqslant 0$$
$$\lambda \geq 0, \quad \mu \geqq 0$$

$$(\text{II}) \qquad Az < 0, \qquad Bz \leqq 0, \qquad Cz = 0$$

Proof. It is easy to shows that (I) and (II) cannot hold simultaneously. Now, Let us assume that (I) has no solution. It is equivalent to the inconsistency of the following system

$$\lambda^t A + \mu^t B + \nu^t C + \xi 0 = 0$$
$$\lambda^t a + \mu^t b + \nu^t c - \xi = 0$$
$$\lambda \geq 0, \quad \mu \geqq 0, \quad \xi \geqslant 0$$

By the Motzkin theorem of the alternative (See [34]), there exist z and $t \leqslant 0$ satisfying

$$Az + a \cdot t < 0, \qquad Bz + b \cdot t \leqq 0, \qquad Cz + c \cdot t = 0. \tag{29}$$

Since system (Υ) has no solution, we have $t = 0$. So, (II) has a solution z. □

Theorem 8 *Let $k \in \{0,1\}$. If $\bar{x}$ is a local (weak) efficient solution to problem (MOP), then for each critical direction $d \in \mathbb{K}$ there exist $\lambda \in \mathbb{R}^{\ell}_+$, $\mu \in \mathbb{R}^{p}_+$, and $\nu \in \mathbb{R}^q$ not all zero such that*

$$\sum_{l=1}^{\ell} \lambda_l \nabla f_l(\bar{x}) + \sum_{i=1}^{p} \mu_i \nabla g_i(\bar{x}) + \sum_{j=1}^{q} \nu_j \nabla h_j(\bar{x}) = 0 \tag{30a}$$

$$k \left(\sum_{l=1}^{\ell} \lambda_l \nabla^2 f_l(\bar{x}) + \sum_{i=1}^{p} \mu_i \nabla^2 g_i(\bar{x}) + \sum_{j=1}^{q} \nu_j \nabla^2 h_j(\bar{x}) \right)(d,d) \geqslant 0 \tag{30b}$$

$$\mu_j g_j(\bar{x}) = 0 \quad \text{for each} \quad j \in I_g \tag{30c}$$

$$\lambda_l \nabla f_l(\bar{x})d = 0 \quad \text{for each} \quad l \in I_f \quad \text{and} \quad \mu_i \nabla g_i(\bar{x})d = 0 \quad \text{for each} \quad i \in I_g \tag{30d}$$

Proof. If $\{\nabla h_j(\bar{x}), \; j \in I_h\}$ are linearly dependent, then one can find vectors $\lambda \geqq 0$, $\mu \geqq 0$ and ν not all zero such that conditions (30) hold. Now suppose that $\{\nabla h_j(\bar{x}), \; j \in I_h\}$ are linearly independent. Then, for $d \in \mathbb{K}$, system (27) has no solution z. Let us assume that system (30) does not hold. By Lemma 5, there exits z such that

$$\nabla f_l(\bar{x})z < 0, \quad l \in I_f(\bar{x}, d); \quad \nabla g_i(\bar{x})z < 0, \quad i \in I_g(\bar{x}, d); \quad \nabla h(\bar{x})z = 0.$$

On the other hand,

$$\nabla f_l(\bar{x})d = 0, \quad l \in I_f(\bar{x}, d); \quad \nabla f_l(\bar{x})d < 0, \quad l \notin I_f(\bar{x}, d)$$
$$\nabla g_i(\bar{x})d = 0, \quad i \in I_g(\bar{x}, d); \quad \nabla g_i(\bar{x})d < 0, \quad i \notin I_g(\bar{x}, d); \quad \nabla h(\bar{x})d = 0$$

because d is a critical direction. Thus, it follows that

$$\nabla f(\bar{x})(d + t\,z) < 0, \quad \nabla g_i(\bar{x})(d + t\,z) < 0, \quad i \in I(\bar{x}); \quad \nabla h(\bar{x})(d + t\,z) = 0$$

for any sufficiently small $t > 0$, which contradicts the first order necessary conditions for efficiency. $\qquad\qquad\qquad\qquad\qquad\qquad\qquad\qquad\qquad\qquad\qquad\qquad\qquad\qquad\qquad\square$

Remark 4 *When $k = 0$, even if Theorem 8 states the characterization of local (weak) efficient solutions in terms of first order derivatives only, it requires second-order differentiability which is the price to be paid to obtain additional optimality condition (30d).*

In scalar objective optimization, the difference between the Fritz John and the Kuhn-Tucker conditions is that the multiplier (λ) of the objective function is assumed to satisfy one of the following conditions, in the later case :

(i) $\lambda \geqslant 0$, $\lambda \neq 0$ (See, for example, [28])

(ii) $\lambda > 0$,   (See, for example, [6])

It is clear that, in scalar objective optimization, conditions (i) and (ii) are equivalent. However, in multiobjective optimization, the equivalence between (i) and (ii) is not conserved. So, we are in position to state two types of Kuhn-Tucker conditions in multiobjective optimization.

4.2   Kuhn-Tucker type 1 necessary optimality conditions

It should be remarked in Theorem 8 that there is no guarantee that $\lambda_l \neq 0$ for at least one $l \in I_f$. In cases where $\lambda = 0$, the objective function do not play any role in the necessary conditions of efficiency. In order to avoid this undesirable situation, some assumptions on the feasible region have to be made.

Definition 9 *The kth Abadie second-order constraint qualification (ASOCQ)$_k$ holds at $\bar{x} \in S$ in the direction $d \in \mathbb{R}^n$ iff*

$$T_k''(S, \bar{x})(d) = L_k''(S, \bar{x})(d). \tag{31}$$

If (ASOCQ)$_k$ holds at $\bar{x} \in S$ in every direction $d \in \mathbb{R}^n$ we said that (ASOCQ)$_k$ holds at $\bar{x} \in S$.

The kth Guignard second-order constraint qualification (GSOCQ)$_k$ holds at $\bar{x} \in S$ in the direction $d \in \mathbb{R}^n$ iff

$$\overline{\text{co}}\,[T_k''(S, \bar{x})(d)] = L_k''(S, \bar{x})(d). \tag{32}$$

It may be noted that, in the special case $d = 0$, conditions (31) and (32) are reduced to (first order) Abadie ([1]) and Guignard ([24]) constraint qualifications.

Theorem 9 *Let $k \in \{0, 1\}$. Suppose that (ASOCQ)$_k$ holds at $\bar{x} \in S$ in the critical direction $d \in \mathbb{K}$. If $\bar{x}$ is a local (weak) efficient solution to problem (MOP), then the system*

$$\nabla f_l(\bar{x})z + k\nabla^2 f_l(\bar{x})(d, d)) < 0, \quad l \in I_f(\bar{x}, d) \tag{33a}$$

$$\nabla g_i(\bar{x})z + k\nabla^2 g_i(\bar{x})(d, d)) \leqslant 0, \quad i \in I_g(\bar{x}, d) \tag{33b}$$

$$\nabla h_j(\bar{x})z + k\nabla^2 h_j(\bar{x})(d, d)) = 0, \quad j \in I_h \tag{33c}$$

has no solution z in $\mathbb{R}^n$.

   Proof.   It follows from Corollary 2 and condition (31).   □

Theorem 10 *Let $k \in \{0, 1\}$. Suppose that (ASOCQ)$_k$ holds at $\bar{x} \in S$ in the critical direction $d \in \mathbb{K}$. If $\bar{x}$ is a local (weak) efficient solution to problem (MOP), then there exist $\lambda \in \mathbb{R}_+^\ell$, $\mu \in \mathbb{R}_+^p$, and $\nu \in \mathbb{R}_+^q$ with $\lambda \neq 0$ such that conditions (30a)–(30d) of Theorem 8 hold.*

   Proof.   The proof follows from Lemma 5 applied to system (33).   □

   Now, we consider the scalar optimization problems and we refine Theorems 9 and 10 for such problems.

Theorem 11 *When $\ell = 1$, Theorems 9 and 10 still hold if the second-order constraint qualification (ASOCQ)$_k$ is replaced by the weaker second-order constraint qualification (GSOCQ)$_k$.*

   Proof.   It follows on the lines of Theorem 9 and Theorem 10, it suffices to take into account Theorem 4 instead Corollary 2.   □

4.3   Kuhn-Tucker type 2 necessary optimality conditions

It should be remarked in Theorem 10 that there is no guarantee that $\lambda > 0$ even if constraint qualifications hold. In cases where $\lambda_l = 0$ for some $l \in I_f$, the corresponding objective function has no role in the necessary conditions of efficiency. In order to avoid this undesirable situation, some assumptions on the problem have to be made.

Definition 10 *The kth Guignard second-order regularity condition $(GSORQ)_k$ holds at $\bar{x} \in S$ in the direction $d \in \mathbb{R}^n$ iff*

$$\bigcap_{i=1}^{\ell} \overline{\text{co}} \, [T_k''(Q^i, \bar{x})(d)] = L_k''(Q, \bar{x})(d). \tag{34}$$

The term "regularity condition" is used instead of "constraint qualifications" because both the objective and the constraint functions are involved.

Theorem 12 *Let $k \in \{0, 1\}$. Suppose that $(GSORQ)_k$ holds at $\bar{x} \in S$ in the critical direction $d \in \mathbb{K}$. If $\bar{x}$ is a (local) efficient solution to problem (MOP), then the system*

$$\nabla f_l(\bar{x})z + k\nabla^2 f_l(\bar{x})(d, d)) \leqslant 0, \quad l \in I_f(\bar{x}, d) \tag{35a}$$

$$\nabla f_s(\bar{x})z + k\nabla^2 f_s(\bar{x})(d, d)) < 0, \quad \text{for at least one } s \tag{35b}$$

$$\nabla g_i(\bar{x})z + k\nabla^2 g_i(\bar{x})(d, d)) \leqslant 0, \quad i \in I_g(\bar{x}, d) \tag{35c}$$

$$\nabla h_j(\bar{x})z + k\nabla^2 h_j(\bar{x})(d, d)) = 0, \quad j \in I_h \tag{35d}$$

has no solution z in $\mathbb{R}^n$.

    Proof.   It follows from Theorem 5 and condition (34).   □
    In order to obtain multiplier rules, the following lemma will be useful.

Lemma 6 *Let A, B and C be given real matrices, with $A \neq 0$. Let a, b and c be real vectors. If the system*

$$(\Theta) \quad\quad Ax + a \leq 0, \quad\quad Bx + b \leqq 0, \quad\quad Cx + c = 0$$

has no solution x in $\mathbb{R}^n$, then exactly one of the two following systems (I) and (II) has a solution:

$$\text{(I)} \quad\quad \lambda^t A + \mu^t B + \nu^t C = 0$$
$$\lambda^t a + \mu^t b + \nu^t c \geqslant 0$$
$$\lambda > 0, \quad \mu \geqq 0$$

$$\text{(II)} \quad\quad Az \leq 0, \quad\quad Bz \leqq 0, \quad\quad Cz = 0$$

    Proof. The proof is similar to the proof of Lemma 5.   □

Theorem 13 *Let $k \in \{0, 1\}$. Suppose that $(GSORQ)_k$ holds at $\bar{x} \in S$ in the critical direction $d \in \mathbb{K}$. If $\bar{x}$ is a (local) efficient solution to problem (MOP), then there exist $\lambda \in \mathbb{R}_+^\ell$, $\mu \in \mathbb{R}_+^p$, and $\nu \in \mathbb{R}_+^q$ with $\lambda_l > 0$ for each $l \in I_f(\bar{x}, d)$ such that conditions (30a)–(30d) of Theorem 8 hold.*

    Proof. The proof follows from Lemma 6 applied to system (35). $\square$

A similar result was given in [2], in which only critical directions satisfying $\nabla f(\bar{x})d = 0$ are considered, and improves in [9] where the authors relax the regularity condition considered in Theorem 3.2 of [2] and give optimality conditions for each critical direction. Our results coincide with Theorem 5.5 in [9] in the case $k = 1$, but the case $k = 0$ is new.

It is worth noting that Theorems 12 and 13 embrace also the first order optimality conditions given by Maeda ([32]), which can be obtained just considering the particular case $d = 0$.

5   John and Kuhn-Tucker sufficient conditions

A first order sufficient condition for (MOP) is that the following system has no nonzero solution y:

$$\nabla f(\bar{x})d \leq 0, \quad \nabla g_E(\bar{x})d \leq 0, \quad \nabla h(\bar{x})d = 0 \tag{36}$$

Now, we shall define three kinds of critical types for problem (MOP).

1. The gap between (36) and the first order Kuhn-Tucker Type 2 necessary condition is caused by the following directions:

$$\nabla f(\bar{x})d = 0, \quad \nabla g_E(\bar{x})d \leq 0, \quad \nabla h(\bar{x})d = 0 \tag{37}$$

    A direction d which satisfies (37) is called a Kuhn-Tucker type 2 critical direction.

2. The gap between (36) and the first order Kuhn-Tucker Type 2 necessary condition is caused by the following directions:

$$\nabla f_l(\bar{x})d = 0 \text{ for at least one } l, \quad \nabla g_E(\bar{x})d \leq 0, \quad \nabla h(\bar{x})d = 0 \tag{38}$$

    A direction d which satisfies (38) is called a Kuhn-Tucker type 1 critical direction.

3. The gap between (36) and the first order John Type necessary condition is caused by (38) and the following directions:

$$\nabla f(\bar{x})d \leq 0, \quad \nabla g_j(\bar{x})d = 0 \text{ for at least one } j \in E, \quad \nabla h(\bar{x})d = 0 \tag{39}$$

    A direction d which satisfies (38) or (39) is called a John type critical direction.

The sets of Kuhn-Tucker type 2 critical directions, Kuhn-Tucker type 1 critical direction, and John type critical directions will be denoted respectively by C_{KT2}, C_{KT1}, C_{FJ}. It is obvious that

$$C_{KT2} \subset C_{KT1} \subset C_{FJ}$$

In order to obtain sufficient optimality conditions we introduce the notions of prequasiinvex [40], weak prequasiinvex [4], and prequasilinear functions

Definition 11 *We say that f is prequasiinvex at $\bar{x} \in X$ with respect to η if X is invex at $\bar{x}$ with respect to η and, for each $x \in X$,*

$$f(x) \leqq f(\bar{x}) \Rightarrow f(\bar{x} + \lambda \eta(x, \bar{x})) \leqq f(\bar{x}), \quad 0 < \lambda \leqslant 0. \tag{40}$$

Definition 12 *We say that f is weak prequasiinvex at $\bar{x} \in X$ with respect to η if X is invex at $\bar{x}$ with respect to η and, for each $x \in X$,*

$$f(x) \leq f(\bar{x}) \Rightarrow f(\bar{x} + \lambda \eta(x, \bar{x})) \leqq f(\bar{x}), \quad 0 < \lambda \leqslant 0. \tag{41}$$

Definition 13 *We say that f is prequasilinear at $\bar{x} \in X$ with respect to η if X is invex at $\bar{x}$ with respect to η and, for each $x \in X$,*

$$f(x) = f(\bar{x}) \Rightarrow f(\bar{x} + \lambda \eta(x, \bar{x})) = f(\bar{x}), \quad 0 < \lambda \leqslant 0. \tag{42}$$

5.1   John type sufficient optimality conditions

Theorem 14 *Suppose that f is weak prequasiinvex, g is prequasiinvex, h is prequasilinear with respect to same η, and all are twice continuously differentiable at $\bar{x} \in S$. Further suppose that $\eta(x, y) \neq 0$ for all $x \neq y$. If for each nonzero $y \in C_{FJ}$, there exist $\lambda \in \mathbb{R}^{\ell}$, $\mu \in \mathbb{R}^p$ and $\nu \in \mathbb{R}^q$ satisfying*

$$\sum_{l=1}^{\ell} \lambda_l \nabla f_l(\bar{x}) + \sum_{i=1}^{p} \mu_i \nabla g_i(\bar{x}) + \sum_{j=1}^{q} \nu_j \nabla h_j(\bar{x}) = 0 \tag{43a}$$

$$k\left(\sum_{l=1}^{\ell} \lambda_l \nabla^2 f_l(\bar{x}) + \sum_{i=1}^{p} \mu_i \nabla^2 g_i(\bar{x}) + \sum_{j=1}^{q} \nu_j \nabla^2 h_j(\bar{x}) \right)(d, d) > 0 \tag{43b}$$

$$\mu_j g_j(\bar{x}) = 0 \quad \text{for each} \quad j \in I_g \tag{43c}$$

$$(\lambda, \mu) \geq 0 \tag{43d}$$

then $\bar{x}$ is (weak) Pareto minimal for (MOP).

Proof.   Assume that, for each critical direction $d \neq 0$, there exit $\lambda \in \mathbb{R}^{\ell}$, $\mu \in \mathbb{R}^p$ and $\nu \in \mathbb{R}^q$ such that (30) hold and that $\bar{x}$ is not Pareto minimal for (MOP). Then, there is $x \in S$ such that

$$f(x) \leq f(\bar{x}), \quad g_E(x) \leqq g_E(\bar{x}), \quad h(\bar{x}) = 0. \tag{44}$$

Using the invexity of f, g_E and h, and (44), we obtain

$$\nabla f(\bar{x})\eta(x,\bar{x}) \leqq 0, \quad \nabla g_E(\bar{x})\eta(x,\bar{x}) \leqq 0, \quad \nabla h(\bar{x})\eta(x,\bar{x}) = 0 \qquad (45)$$

Two cases are to be considered:

Case 1. If $\nabla f(\bar{x})\eta(x,\bar{x}) < 0$ and $\nabla g_E(\bar{x})\eta(x,\bar{x}) < 0$, then $d = \eta(x,\bar{x})$ is solution to the following system

$$\nabla f(\bar{x})d < 0, \quad \nabla g_E(\bar{x})d < 0, \quad \nabla h(\bar{x})d = 0.$$

By the Motzkin theorem of the alternative, the following system

$$\lambda \nabla f(\bar{x}) + \mu \nabla g(\bar{x}) + \nu \nabla h(\bar{x}) = 0,$$

$$\mu g(\bar{x}) = 0,$$

$$(\lambda, \mu) \geq 0,$$

is inconsistent. Then, contradicting (43a), (43c) and (43d).

Case 2. If $\nabla f_r(\bar{x})\eta(x,\bar{x}) = 0$, for at least one r or $\nabla g_j(\bar{x})\eta(x,\bar{x}) = 0$, for at least one $j \in E$, then $d = \eta(x,\bar{x})$ is a nonzero critical direction. Let $0 < t \leqslant 1$, from the weak prequasiinvexity of f, we get

$$f(\bar{x}+td) - f(\bar{x}) = t\nabla f(\bar{x})d + t^2/2\nabla^2 f(\bar{x})(d,d) + o(t^2) \leqq 0$$

where $o(t^2)$ is a vector satisfying $\|o(t^2)\|/t^2$, hence

$$\nabla f(\bar{x})y + t/2[\nabla^2 f(\bar{x})(y,y) + o(t^2)/t^2] \leqq 0. \qquad (46)$$

Similarly,

$$\nabla g_E(\bar{x})y + t/2[\nabla^2 g_E(\bar{x})(y,y) + o(t^2)/t^2] \leqq 0. \qquad (47)$$
$$\nabla h(\bar{x})y + t/2[\nabla^2 h(\bar{x})(y,y) + o(t^2)/t^2] = 0. \qquad (48)$$

Multiplying (46), (47), (48) by λ, μ, ν and summing, we get

$$(\lambda \nabla f(\bar{x}) + \mu \nabla g_E(\bar{x}) + \nu \nabla f(\bar{x}))\, d$$

$$+\frac{t}{2}\left(\sum_{l=1}^{\ell}\lambda_l \nabla^2 f_l(\bar{x}) + \sum_{i=1}^{p}\mu_i \nabla^2 g_i(\bar{x}) + \sum_{j=1}^{p}\nu_j \nabla^2 h_j(\bar{x})\right)(d,d) + \frac{t}{2}o(t^2)/t^2 \leqq 0.$$

From expressions (43a), (43c), and $t > 0$, we get

$$\left(\sum_{l=1}^{\ell}\lambda_l \nabla^2 f_l(\bar{x}) + \sum_{i=1}^{p}\mu_i \nabla^2 g_i(\bar{x}) + \sum_{j=1}^{q}\mu_j \nabla^2 h_j(\bar{x})\right)(d,d) + o(t^2)/t^2 \leqq 0.$$

Letting $t \to 0$, we obtain

$$\left(\sum_{l=1}^{\ell} \lambda_l \nabla^2 f_l(\bar{x}) + \sum_{i=1}^{p} \mu_i \nabla^2 g_i(\bar{x}) + \sum_{j=1}^{q} \mu_j \nabla^2 h_j(\bar{x}) \right)(d, d) \leqq 0, \tag{49}$$

contradicting (43b). Hence, $\bar{x}$ is Pareto minimal for (MOP). □

5.2   Kuhn-Tucker type 1 sufficient optimality conditions

Theorem 15 *Suppose that f is weak prequasiinvex, g is prequasiinvex, h is prequasilinear with respect to same η, and all are twice continuously differentiable at $\bar{x} \in S$. Further suppose that $\eta(x, y) \neq 0$ for all $x \neq y$. If for each nonzero $y \in C_{KT1}$, there exist $\lambda \in \mathbb{R}^{\ell}$, $\mu \in \mathbb{R}^p$ and $\nu \in \mathbb{R}^q$ satisfying*

$$\sum_{l=1}^{\ell} \lambda_l \nabla f_l(\bar{x}) + \sum_{i=1}^{p} \mu_i \nabla g_i(\bar{x}) + \sum_{j=1}^{q} \nu_j \nabla h_j(\bar{x}) = 0 \tag{50a}$$

$$k \left(\sum_{l=1}^{\ell} \lambda_l \nabla^2 f_l(\bar{x}) + \sum_{i=1}^{p} \mu_i \nabla^2 g_i(\bar{x}) + \sum_{j=1}^{q} \nu_j \nabla^2 h_j(\bar{x}) \right)(d, d) > 0 \tag{50b}$$

$$\mu_j g_j(\bar{x}) = 0 \quad \text{for each} \quad j \in I_g \tag{50c}$$

$$\lambda \geq 0, \quad \mu \geqq 0 \tag{50d}$$

then $\bar{x}$ is (weak) Pareto minimal for (MOP).

Proof.   Assume that, for each critical direction $d \neq 0$, there exit $\lambda \in \mathbb{R}^{\ell}$, $\mu \in \mathbb{R}^p$ and $\nu \in \mathbb{R}^q$ such that (50) hold and that $\bar{x}$ is not Pareto minimal for (MOP). Then, there is $x \in S$ such that (45) hold. Two cases are to be considered:

Case 1.   If $\nabla f(\bar{x}) \eta(x, \bar{x}) < 0$, then $d = \eta(x, \bar{x})$ is solution to the following system

$$\nabla f(\bar{x})d < 0, \quad \nabla g_E(\bar{x})d \leqq 0, \quad \nabla h(\bar{x})d = 0.$$

By the Motzkin theorem of the alternative, the following system

$$\lambda \nabla f(\bar{x}) + \mu \nabla g(\bar{x}) + \nu \nabla h(\bar{x}) = 0,$$

$$\mu g(\bar{x}) = 0,$$

$$\lambda \geq 0, \quad \mu \geqq 0,$$

is inconsistent. Then, contradicting (50a), (50c) and (50d).

Case 2. If $\nabla f_r(\bar{x})\eta(x,\bar{x}) = 0$, for at least one r, then $d = \eta(x,\bar{x})$ is a nonzero Kuhn-Tucker type 1 critical direction. From the weak prequasiinvexity of f, we get (46). Similarly, we obtain (47) and (48). Using the same arguments as in Theorem 14, relation (49) hold which contradicts (50b). This completes the proof. $\qquad\square$

5.3 Kuhn-Tucker type 2 sufficient optimality conditions

Theorem 16 *Suppose that f is weak prequasiinvex, g is prequasiinvex, h is prequasilinear with respect to same η, and all are twice continuously differentiable at $\bar{x} \in S$. Further suppose that $\eta(x,y) \neq 0$ for all $x \neq y$. If for each nonzero $y \in C_{KT2}$, there exist $\lambda \in \mathbb{R}^\ell$, $\mu \in \mathbb{R}^p$ and $\nu \in \mathbb{R}^q$ satisfying*

$$\sum_{l=1}^{\ell} \lambda_l \nabla f_l(\bar{x}) + \sum_{i=1}^{p} \mu_i \nabla g_i(\bar{x}) + \sum_{j=1}^{q} \nu_j \nabla h_j(\bar{x}) = 0 \tag{51a}$$

$$k\left(\sum_{l=1}^{\ell} \lambda_l \nabla^2 f_l(\bar{x}) + \sum_{i=1}^{p} \mu_i \nabla^2 g_i(\bar{x}) + \sum_{j=1}^{q} \nu_j \nabla^2 h_j(\bar{x}) \right)(d,d) > 0 \tag{51b}$$

$$\mu_j g_j(\bar{x}) = 0 \quad \text{for each} \ \ j \in I_g \tag{51c}$$

$$\lambda > 0, \quad \mu \geqq 0 \tag{51d}$$

then $\bar{x}$ is (weak) Pareto minimal for (MOP).

Proof. Assume that, for each critical direction $d \neq 0$, there exit $\lambda \in \mathbb{R}^\ell$, $\mu \in \mathbb{R}^p$ and $\nu \in \mathbb{R}^q$ such that (50) hold and that $\bar{x}$ is not Pareto minimal for (MOP). Then, there is $x \in S$ such that (45) hold. Two cases are to be considered:

Case 1. If $\nabla f(\bar{x})\eta(x,\bar{x}) \leq 0$, then $d = \eta(x,\bar{x})$ is solution to the following system

$$\nabla f(\bar{x})d \leq 0, \quad \nabla g_E(\bar{x})d \leqq 0, \quad \nabla h(\bar{x})d = 0.$$

By the Tucker theorem of the alternative, the following system

$$\lambda \nabla f(\bar{x}) + \mu \nabla g(\bar{x}) + \nu \nabla h(\bar{x}) = 0,$$
$$\mu g(\bar{x}) = 0,$$
$$\lambda > 0, \quad \mu \geqq 0,$$

is inconsistent. Then, contradicting (51a), (51c) and (51d).

Case 2. If $\nabla f(\bar{x})\eta(x,\bar{x}) = 0$, then $d = \eta(x,\bar{x})$ is a nonzero Kuhn-Tucker type 2 critical direction. From the weak prequasiinvexity of f, we get (46). Similarly, we obtain (47) and (48). Using the same arguments as in Theorem 14, relation (49) hold which contradicts (51b). This completes the proof. $\qquad\square$

6   General second-order necessary conditions

In this section we prove very general second-order necessary conditions by generalizing the main Theorem 1 to the case of any ordering convex pointed cone. Since these conditions do not require differentiability assumption, they are basic to deducing second-order necessary conditions of problems with data in a larger class of functions.

Let $Q \subset \mathbb{R}^{\ell}$ be a convex, closed, and pointed $(Q \cap (-Q) = \{0\})$ cone with apex 0 and nonempty interior(int $Q \neq \varnothing$). This cone Q is called an ordering cone. For x, $y \in \mathbb{R}^{\ell}$, by $x \leqq_Q y$, we mean $x - y \in Q$; by $x \leq_Q y$, we mean $x - y \in Q \setminus \{0\}$ and by $x <_Q y$, we mean $x - y \in \text{int } Q$. The concepts of efficiency with respect to cone Q are defined as in definitions 1 and 2 with the lower index "$_Q$": $\leq_Q$ and $<_Q$.

In order to prove necessary conditions for efficiency with respect to Q, we need the lemma below.

Lemma 7 *Let $Q \subset \mathbb{R}^{\ell}$ be a convex set, and let $d \in \overline{Q}$ and $z \in \mathbb{R}^{\ell}$. If there is a $t > 0$ such that $d + tz \in \text{int } Q$, then there exists a $\delta > 0$ such that $d + \alpha B(z, \delta) \subset \text{int } Q$ for each $\alpha \in (0, t]$.*

Proof.   It follows immediately from Theorem 2.2.2 in [6].   □

Theorem 17 (Fondamental Theorem.) *If $\bar{y}$ is a local (weak) efficient point of $Z \subset \mathbb{R}^{\ell}$, then for every $k \in \{0, 1\}$ the following holds*

$$T_k''(Z, \bar{y}) \cap \Omega_Q = \varnothing, \tag{52}$$

where $\Omega_Q = \left\{ (d, z) \in \mathbb{R}^{\ell} \times \mathbb{R}^{\ell} \mid d \in \text{int } Q \;\; or \;\; [d \in \text{Fr}(Q) \; and \; z \in \text{int cone}(Q - d)] \right\}$

Proof.   Suppose to the contrary that (52) does not hold, that is there exist $(d, z) \in \Omega_Q$, $(t_n, r_n) \longrightarrow (0^+, 0^+)$ and $z_n \longrightarrow z$ such that $t_n/r_n \longrightarrow k$ and (6) is verified. If $d \in \text{int } Q$, since $||r_n z_n|| \longrightarrow 0$, one has $d + (1/2)r_n z_n \in \text{int } Q$ so that $y_n - \bar{y} \in \text{int } Q$ when n is sufficiently large which contradicts that $\bar{y}$ is a local (weak) efficient point of Z. If $d \in \text{Fr}(Q)$, hence $z \in \text{int cone}(Q - d)$. Or,

$$\begin{aligned}
\text{int cone}(Q - d) &= \{\lambda(x - d) | \lambda > 0, \; x \in \text{int } Q\} \\
&= \{w \in \mathbb{R}^{\ell} | \exists t > 0 \text{ such that } d + tw \in \text{int } Q\}
\end{aligned}$$

Hence, there exists $t > 0$ such that $d + tz \in \text{int } Q$. By Lemma 7, there exists $\delta > 0$ such that $d + \alpha B(z, \delta) \subset \text{int } Q$, for each $\alpha \in (0, t]$. Since $z_n \longrightarrow z$ and $r_n \longrightarrow 0^+$, one has $z_n \in B(z, \delta)$ and $(1/2)r_n < t$ for n sufficiently large. Therefore, $d + (1/2)r_n z_n \in \text{int } Q$ which implies that $y_n - \bar{y} \in \text{int } Q$ when n is sufficiently large which contradicts again that $\bar{y}$ is a local (weak) efficient point of Z.   □

In the case when $Q = -\mathbb{R}_+^{\ell}$, the set Ω_Q is reduced to Ω. Also, as observed in the proof, Theorem 3 hold when we replace $\mathbb{R}^{\ell}$ with a normed space Y.

Corollary 3 *If $\bar{y}$ is a local (weak) efficient point of $Z \subset \mathbb{R}^\ell$, then for every direction $d \in T(Z, \bar{y}) \cap \operatorname{Fr} Q$ the following holds*

$$T_k''(Z, \bar{y})(d) \cap \Omega_Q(d) = \varnothing, \quad k \in \{0, 1\} \tag{53}$$

where $\Omega_Q(d) = \{z \in \mathbb{R}^\ell \mid (d, z) \in \Omega_Q \}$

Proof. It follows immediately from Theorem 1. ☐

We note that, for the particular case $d = 0$, we find Theorem 3.1 in [18].

As an application of Theorem 3 we prove the following second-order necessary optimality conditions

Theorem 18 *Let f be twice continuously differentiable at $\bar{x} \in S$. If $\bar{x}$ is a local (weak) efficient solution to problem* (MOP), *then the system*

$$\nabla f(\bar{x})d \in \operatorname{int} Q \quad or$$
$$\nabla f(\bar{x})d \in \operatorname{Fr} Q \quad and \quad \nabla f(\bar{x})z + k\nabla^2 f(\bar{x})(d, d) \in \operatorname{int} \operatorname{cone} (Q - \nabla f(\bar{x})d)$$

has no solution (d, z) in $T_k''(S, \bar{x})$ for every $k \in \{0, 1\}$.

Proof. The proof of this theorem is similar to that of Theorem 3 (we replace there Theorem 1 and Ω by Theorem 17 and Ω_Q, respectively). ☐

Now, let $f : \mathbb{R}^n \longrightarrow \mathbb{R}^\ell$ be of class $C^{1,1}$, which means that ∇f is locally Lipschitz. The Clarke generalized Jacobian of ∇f at $\bar{x}$, called the second-order subdifferential of f at $\bar{x}$, is

$$\partial^2 f(\bar{x}) = \overline{\operatorname{co}} \{\lim \nabla^2 f(x_n) : x_n \longrightarrow \bar{x}, \ \nabla^2 f(x_n) \text{ exists}\}$$

For more properties of generalized subdifferentials and their relations, see [17] and [27].

Theorem 19 *Let f be $C^{1,1}$ function and $\bar{x} \in S$. If $\bar{x}$ is a local (weak) efficient solution to problem* (MOP), *then there is $M \in \partial^2 f(\bar{x})$ such that the system*

$$\nabla f(\bar{x})d \in \operatorname{int} Q \quad or$$
$$\nabla f(\bar{x})d \in \operatorname{Fr} Q \quad and \quad \nabla f(\bar{x})z + kM(d, d) \in \operatorname{int} \operatorname{cone} (Q - \nabla f(\bar{x})d)$$

has no solution (d, z) in $T_k''(S, \bar{x})$ for every $k \in \{0, 1\}$.

Proof. Here, the proof of Theorem 3 is valid upto formula (13). By Taylor's expansion, there exist $\varepsilon_n \longrightarrow 0$ such that

$$\begin{aligned}
f(x_n) &= f(\bar{x}) + t_n \nabla f(\bar{x})d + (1/2)t_n r_n \Big[\nabla f(\bar{x})(z_n) \\
&\quad + k_n M_n \left(d + (1/2)r_n z_n, d + (1/2)r_n z_n \right) + 2k_n \varepsilon_n \Big]
\end{aligned}$$

for some $M_n \in \overline{co}\{\partial^2 f(x) : x \in [\bar{x}, x_n]\}$ where $k_n = t_n/r_n \longrightarrow k$. Since the set-valued map $x \longmapsto \partial^2 f(x)$ is upper semicontinuous and convex, compact-valued (see [17] and [22]), we may assume that M_n converges to some $M \in \partial^2 f(\bar{x})$. Then

$$\nabla f(\bar{x})(z_n) + k_n M_n \left(d + (1/2)r_n z_n, d + (1/2)r_n z_n\right) \underset{n \to +\infty}{\longrightarrow} \nabla f(\bar{x})z + kM(d,d)$$

which implies that

$$(\nabla f(\bar{x})d, \nabla f(\bar{x})z + kM(d,d)) \in T_k''(f(S \cap N), f(\bar{x})).$$

Since $f(\bar{x})$ is (weak) efficient point of $f(S \cap N)$, Theorem 17 leads to

$$T_k''(f(S \cap N), f(\bar{x})) \cap \Omega_Q = \varnothing.$$

Hence

$$(\nabla f(\bar{x})d, \nabla f(\bar{x})z + kM(d,d)) \notin \Omega_Q.$$

This complete the proof.   □

This theorem extends Theorem 3.1 in [22] in which the second-order set is used but the asymptotic second-order tangent cone is not considered. On the other hand, Theorem 3.1 in [22] says nothing about (critical) directions d in $T(S, \bar{x})$ such that

$$\nabla f(\bar{x})d \in \mathrm{Fr}\, Q \quad \text{and} \quad \nabla f(\bar{x})d \neq 0.$$

The same happens also with Theorem 3.1 in [23] in which authors give second-order necessary conditions when the objective fonction f is of class C^1 i.e. f is continuously differentiable. Taking account of Theorem 17, if we make appropriate modifications in the statement of Theorem 3.1 in [23], we can prove it following the line of the proof in [23].

References

[1] J. Abadie, On the Kuhn-Tucker Theorem, Nonlinear Programming, Edited by J. Abadie, North Holland, Amsterdam, Netherlands, (1967) 21–36.

[2] B. Aghezzaf, Second-Order Necessary Conditions of the Kuhn-Tucker Type in Multiobjective Programming Problems, Control and Cybernetics 28 (1999) 213–224.

[3] B. Aghezzaf and M. Hachimi, Second-Order Optimality Conditions in Multiobjective Optimization Problems, Journal of Optimization Theory and Applications 102 (1999) 37–50.

[4] B. Aghezzaf and M. Hachimi, Sufficient Optimality Conditions and Duality in Multiobjective Optimization Involving Generalized Convexity, Numerical Functional Analysis and Optimization 22 (2001) 775–788.

[5] B. Aghezzaf and M. Hachimi, On a Gap between Multiobjective Optimization and Scalar Optimization, Journal of Optimization Theory and Applications 109 (2001) 431–435.

[6] M.S. Bazaraa, H.D. Sherali and C.M. Shetty, Nonlinear Programming : Theory and Algorithms, John Wiley & Sons, 2nd Edition, 1993.

[7] A. Ben-Tal, Second-Order and Related Extremality Conditions in Nonlinear Programming, Journal of Optimization Theory and Applications 31 (1980) 143–165.

[8] A. Ben-Tal and J. Zowe, A Unified Theory of First and Second-Order Conditions for Extremum Problems in Topological Vector Spaces, Mathematical Programming Study 19 (1982) 39–76.

[9] G. Bigi and M. Castellani, Second-Order Optimality Conditions for Differentiable Multiobjective Problems, RAIRO - Operations Research 34 (2000) 411–426.

[10] S. Bolintinéanu and M. El Maghri, Second-Order Efficiency Conditions and Sensitivity of Efficient Points, Journal of Optimization Theory and Applications 98 (1998) 569–592.

[11] A. Cambini, S. Komlósi and L. Martein, Recent Developments in Second-Order Necessary Optimality Conditions, Generalized Convexity, Generalized Monoticity, Edited by J.P. Crouzeix, J.E. Martínez-Legaz and M. Volle, Kluwer Academic Publishers, Amsterdam, Netherlands, (1998) 347–356.

[12] A. Cambini, L. Martein and R. Cambini, A New Approach to Second-Order Optimality Conditions in Vector Optimization, Proceeding of the 2nd International conference on multiobjective programming and goal programming, Edited by R.E. Steuer, Springer Verlag, Berlin, Germany, (1997) 219-227.

[13] A. Cambini, L. Martein and M. Vlach, Second-Order Tangent Sets and Optimality Conditions, Research Report, Japan Advanced Studies of Science and Technology, Hokuriku, Japan, 1997.

[14] R. Cambini, Second-Order Optimality Conditions in Multiobjective Programming, Optimization 44 (1998) 139–160.

[15] M. Castellani and M. Pappalardo, Local Second-order Approximations and Applications in Optimization, Optimization 4 (1996) 305–321.

[16] M. Castellani and M. Pappalardo, About a Gap between Multiobjective Optimization and Scalar Optimization, Journal of Optimization Theory and Applications 109 (2001) 437–439.

[17] F.H. Clarke, Optimization and Nonsmmoth Analysis, Wiley, New York, 1983.

[18] H.W. Corley, On Optimality Conditions for Maximizations with Respect to Cones, Journal of Optimization Theory and Applications 46 (1985) 67–78.

[19] H.W. Corley, Optimality Conditions for Maximizations of Set-Valued Functions, Journal of Optimization Theory and Applications 58 (1988) 1–10.

[20] I. Ginchev, A. Guerraggio and M. Rocca, From scalar to vector optimization, Applied Mathematics 51 (2006) 5–36.

[21] I. Ginchev, V.I. Ivanov, Second-order optimality conditions for problems with C^1 data, Journal of Mathematical Analysis and Applications 340 (2008) 646–657.

[22] A. Guerraggio and D.T. Luc, Optimality Conditions for $C^{1,1}$ Constrained Multiobjective Problems, Journal of Optimization Theory and Applications 116 (2003) 117–129.

[23] A. Guerraggio, D.T. Luc and N.B. Minh, Second-Order Optimality Conditions for C^1 Multiobjective Programming Problems, Acta Vietnamica 26 (2001) 257–268.

[24] M. Guignard, Generalized Kuhn-Tucker Conditions for Mathematical Programming problems in a Banach space, SIAM Journal on Control 7 (1987) 232-241.

[25] M. Hachimi, Conditions d'Optimalité et Dualité en Programmation Mathématique Multiobjectif, Thèse de Doctorat, Université Hassan II–Aïn Chock, Maroc, 2000.

[26] M. Hachimi and B. Aghezzaf, New Results on Second-Order Optimality Conditions in Vector Optimization Problems, Journal of Optimization Theory and Applications 135 (2007) 117–133.

[27] J. Jahn, Introduction to the Theory of Nonlinear Optimization, Springer Verlag, Berlin, Germany, 1994.

[28] H. Kawasaki, Second-Order Necessary Conditions of The Kuhn-Tucker Type under New Constraint Qualification, Journal of Optimization Theory and Applications 57 (1988) 253–264.

[29] H.W. Kuhn and F.H. Tucker, Nonlinear Programming, Proceedings of the 2nd Berkeley Symposium on Mathematical Statistics and Probability, Berkeley, California (1951) 481–492.

[30] J.G. Lin, Maximal Vectors and Multiobjective Optimization, Journal of Optimization Theory and Applications 18 (1976) 41–64.

[31] D.T. Luc, Theory of Vector Optimization, Lecture Notes in Economics and Mathematical Systems, Springer Verlag, Berlin, Germany, Vol. 319, 1989.

[32] T. Maeda, Constraint Qualification in Multiobjective Optimization Problems : Differentiable Case, Journal of Optimization Theory and Applications 80 (1994) 483–500.

[33] T. Maeda, Second-Order Conditions for Efficiency in Nonsmooth Multiobjective Optimization Problems, Journal of Optimization Theory and Applications 122 (2004) 521–538.

[34] O.L. Mangasarian, Nonlinear Programming, McGrow Hill, New York, 1969.

[35] K. Pastor, A note on second-order optimality conditions, Nonlinear Analysis 71 (2009) 1964–1969.

[36] J.P. Penot, Optimality Conditions in Mathematical Programming and Composite Optimization, Mathematical Programming 67 (1994) 225–245.

[37] J.P. Penot, Second-Order Conditions for Minimization Problems with Constraints, SIAM Journal on Control and Optimization 37 (1998) 303–318.

[38] Y. Sawaragi, H. Nakayama and T. Tanino, Theory of Multiobjective Optimization, Academic Press, Orlando, Florida, 1985.

[39] C. Singh, Optimality Conditions in Multiobjective Differentiable Programming, Journal of Optimization Theory and Applications 53 (1987) 115–123.

[40] S.K. Suneja, C. Singh and C.R. Bector, Generalization of Preinvex and B-vex Functions, Journal of Optimization Theory and Applications 76 (1993) 577–587.

[41] S. Wang, Second-Order Necessary and Sufficient Conditions in Multiobjective Programming, Numerical Functional Analysis and Optimization 12 (1991) 237–252.

[42] S.Y. Wang and F.M. Yang, A Gap between Multiobjective Optimization and Scalar Optimization, Journal of Optimization Theory and Applications 68 (1991) 389–391.

[43] P.L. Yu, Multiple-Criteria Decision Making: Concepts, Techniques and Extensions, Plenum Press, New York, 1985.

CHAPTER 4

Invex functions and existence of weakly efficient solutions for nonsmooth vector optimization.*

Lucelina Batista Santos[†]   Marko Rojas-Medar[‡]   Gabriel Ruiz-Garzón[§]

Antonio Rufián-Lizana[¶]

Abstract

In this work we study the existence of weakly efficient solutions for some non-smooth and nonconvex vector optimization problems. We consider problems whose objective functions are defined between infinite and finite-dimensional Banach spaces. Our results are stated under hypotheses of generalized convexity and make use of variational-like inequalities.

keywords: Vector Optimization; Nonsmooth Analysis; Invex Functions; Variational-like inequalities.

1 Introduction

The connection between variational inequalities and optimization problems is well known (e.g. [1], [6], [11], [12]) and have been extensively investigated in the recent years by several authors. One of the main works in this direction was done by Giannessi [8], where many existence results for optimization problems were obtained by using of variational inequalities.

For multiobjective optimization problems, Giannessi proved in [9] that there exists an equivalence between efficient solutions of differentiable convex optimization problems and the solutions of a variational inequality of Minty type. He also established similar results for efficient solutions. On the other hand, using subdifferentials, Lee showed in [13]

---

*This work was partially supported by the grant MTM 2007-063432 of the Science and Education Spanish Ministry, CNPq-Brazil and FONDECYT-Chile.

[†]Departamento de Matemática. Universidade Federal do Paraná. CP 19081 CEP, 81531-990 Curitiba, Paran´a, Brazil e-mail: lucelina@ufpr.br

[‡]Universidad Del Bío-Bío. Facultad de Ciencias. Departamento de Ciencias Básicas. Casilla: 447 Av. Andrés Bello S/N Chillán- Chile. e-mail: marko@ueubiobio.cl

[§]Departamento de Estadística e Investigación Operativa. Facultad de Ciencias Sociales y de la Comunicación. Universidad de Cádiz, Spain e-mail: gabriel.ruiz@uca.es

[¶]Departamento de Estadística e Investigación Operativa, Facultad de Matemáticas, Universidad de Sevilla, Sevilla, Spain. e-mail:rufian@us.es

that analogous results are true fore nonsmooth convex problems defined between finite-dimensional spaces.

For some nonconvex differentiable vector problems defined between infinite-dimensional Banach spaces, Chen and Craven [4] proved the equivalence of weakly efficient solutions and the solutions of a certain variational-like inequality. Using this characterization they proved an existence result for weakly efficient solutions.

In this work we consider the following two problems:

1. **The infinite-dimensional problem:**

$$\left. \begin{array}{l} \text{Minimize } f(x) \\ \text{subject to: } x \in K \end{array} \right\} \tag{P1}$$

where X and Y are two Banach spaces, $f : X \to Y$ is a given function and K is a nonempty subset of X and Y is (partially) ordered by a convex cone P.

2. **The finite-dimensional problem:**

$$\left. \begin{array}{l} \text{Minimize } f(x) := (f_1(x), ..., f_p(x)) \\ \text{subject to: } x \in X \end{array} \right\} \tag{P2}$$

where $f_i : \mathbb{R}^n \to \mathbb{R}$ $(i = 1, .., p)$ are given functions and X is a nonempty subset of $\mathbb{R}^n$.

For both problems, by "minimize" we mean "find the weakly efficient solutions of the problem". Our objective is to solve the problem (P1) and (P2) without assuming hypotheses of differentiability on the data of the problems. Our results extend the corresponding results of Chen and Craven [4] and Lee [13].

This Chapter is organized as follows: in Section 2, we fix the notation and we recall some facts from nonsmooth analysis. In Section 3, we consider the problem (P1) and we establish our existence result for this problem and in the Section 3 we will do the same for the problem (P2).

2   Preliminaries

Let X and Y be two real Banach spaces. We will denote by $|| \cdot ||$ the norm in Y. Let K be a nonempty subset of X and $P \subset Y$ a pointed convex cone (i.e., P is a convex cone that satisfies $P \cap (-P) = \{0\}$) such that $\text{int } P \neq \emptyset$. Also, let $f : X \to Y$ be a given function. We consider the problem (P1) given in the previous section. The notion of optimality that we consider here is the *weak efficiency*. We say that $x_0 \in K$ is a **weakly efficient solution** of (P1) if

$$f(x) - f(x_0) \notin - \text{int } P, \forall x \in K.$$

In particular, for the problem (P2), the definition of weakly efficient solution will be done by taking $P = \mathbb{R}_+^p$ in the previous definition- that is, $x_0 \in X$ is a weakly efficient solution of (P2) if does not exist $x \in X$ such that

$$f_i(x) < f_i(x_0), \forall i = 1, ..., p.$$

Now, we recall some notions and results from nonsmooth analysis. Let ϕ be a locally Lipschitz function from a Banach space X into $\mathbb{R}$. The *Clarke generalized directional derivative* of ϕ at a point $\overline{x} \in X$ in the direction $d \in X$, denoted by $\phi^0(\overline{x}; d)$ is given by:

$$\phi^0(\overline{x}; d) = \limsup_{\substack{y \to \overline{x} \\ t \downarrow 0}} \frac{\phi(y + td) - \phi(y)}{t}$$

and the *Clarke generalized gradient* of ϕ at $\overline{x}$ is given by

$$\partial\phi(\overline{x}) = \{x^* \in X^* : \phi^0(\overline{x}; d) \geq \langle x^*, d \rangle, \forall d \in X\}$$

where X^* denotes the topological dual of X and $\langle \cdot, \cdot \rangle$ is the canonical bilinear form between X^* and X.

The next proposition establishes some properties of the generalized directional derivative and the generalized gradient of Clarke.

Proposition 1 *Let $f : \Omega \to \mathbb{R}$ be a locally Lipschitz function with Lipschitz constant k. Then the following assertions are true:*

1. *The function $v \mapsto f^0(x; v)$ is finite, sublinear and satisfies*

$$|f^0(x; v)| \leq k\|v\|;$$

2. *For each $x \in \Omega$, $\partial f(x)$ is a w^*-compact and nonempty subset of X^*. Furthermore, $\|\xi\| \leq k$, for all $\xi \in \partial f(x)$ where*

$$\|\xi\| = \sup_{\substack{x \in X \\ \|x\| \leq 1}} \langle \xi, x \rangle, \xi \in X^*;$$

3. *For each $v \in X$, we have*

$$f^0(x; v) = \max\{\langle \xi, v \rangle : \xi \in \partial f(x)\};$$

4. *One has $\xi \in \partial f(x)$ if and only if $f^0(x; v) \geq \langle \xi, v \rangle$, for each $v \in X$;*

5. *The function $(x, v) \mapsto f^0(x; v)$ is upper semicontinuous.*

A function $f : \Omega \to \mathbb{R}$ is called *regular* (or Clarke regular) at $x \in \Omega$ if

(i) For each $v \in X$ there exists the usual directional derivative of f at x, in the direction v, denoted by $f'(x; v)$;

(ii) For all $v \in X$, $f^0(x; .v) = f'(x; v)$.

Furthermore, if f is regular at x for each $x \in \Omega$ we say that f is regular in Ω.

Proposition 2 *These following assertions are true:*

(a) *If f_i are regular at $x \in \Omega$, then $\partial(\sum_{i=1}^n f_i)(x) = \sum_{i=1}^n \partial f_i(x)$. (The inclusion $\subset$ is true, without the regularity assumption);*

(b) *If f_i is convex and Lipschitz near $x \in \Omega$ then f is regular at x;*

(c) *If f is continuously differentiable at $x \in \Omega$ then $\partial f(x) = \{f'(x)\}$, where f' is usual derivative.*

We note that if C is a nonempty subset of X, then the distance function $d_C : X \to \mathbb{R}$, defined by $d_C(x) = \inf\{\|x - c\| : c \in C\}$ is not differentiable but it is (globally) Lipschitz.

Let C be a nonempty subset of X and $x \in C$. We say that $v \in X$ is a tangent vector to C at x if $d_C^0(x; v) = 0$. We denote by $T_C(x)$ the set of all tangent vector to C at x and $T_C(x)$ is called *tangent cone* of C at x. The *normal cone* of C at x is defined by

$$N_C(x) = \{\xi \in X^* : \langle \xi, v \rangle \leq 0, \forall v \in T_C(x)\}.$$

It can be proved that, for each $x \in C$, $T_C(x)$ is a convex cone, closed in X and $N_C(x)$ is a convex cone, w*-closed in X^*. The following Proposition establishes a necessary condition for optimality.

Proposition 3 *Let $f : \Omega \to \mathbb{R}$ be a locally Lipschitz function and let x^* be a minimum of f in $C \subset \Omega$. Then,*

$$0 \in \partial f(x^*) + N_C(x^*). \tag{1}$$

Note that the condition (1) is equivalent to

$$f^0(x^*; v) \geq 0, \forall v \in T_C(x^*). \tag{2}$$

For more details on nonsmooth analysis, we refer the reader the book of Clarke [5].

Next, we recall the definition of *strongly compactly Lipschitz function*, which is very important for the analysis of the infinite-dimensional problem (P1).

Definition 4 (Thibault [20]) *A function $h : X \to Y$ is said to be **compactly Lipschitz** at $\overline{x} \in X$ if there exists a multifunction (that is, a point-to-set map) $R : X \longrightarrow \text{Comp}(Y)$ where $\text{Comp}(Y)$ denotes the set of all compact subsets of Y and there exists a function $r : X \times X \to \mathbb{R}_+$ satisfying*

(i) $\lim_{\substack{x \to \overline{x} \\ d \to 0}} r(x, d) = 0;$

(ii) There exists $\alpha \geq 0$ such that

$$t^{-1}[h(x + td) - h(x)] \in R(d) + ||d|| r(x, t) B_Y$$

for all $x \in \overline{x} + \alpha B_Y$ and $t \in (0, \alpha)$, where B_Y denotes the closed unit ball around the origin of Y;

(iii) $R(0) = \{0\}$ and R is an upper semicontinuous multifunction.

Moreover, we say that h is a strongly compactly Lipschitz function if it is strongly compactly Lipschitz at each $x \in X$.

If Y has finite dimension, then h is strongly compactly Lipschitz at $\overline{x}$ if and only if it is locally Lipschitz near $\overline{x}$. If h is strongly compactly Lipschitz, then $(u^* \circ h)(x) = \langle u^*, h(x) \rangle$ is locally Lipschitz, for all $u \in Y^*$. This fact is very important because it allows us to extend some results of the nonsmooth analysis to functions defined between infinite-dimensional spaces. For more details about strongly Lipschitz functions we refer the reader to [20].

3　The infinite-dimensional problem (P1)

We recall some definitions of generalized convexity for functions defined between Banch spaces. Given a cone $P \subset Y$, the dual cone of P is defined by

$$P^* = \{\xi \in Y^* : \langle \xi, x \rangle \geq 0, \forall x \in P\}.$$

It can be proved that, if P is a convex cone with int $P \neq \emptyset$, then $\langle \xi, v \rangle > 0$, for all $v \in$ int P. See [10]. We will need the following concepts of generalized convexity:

Definition 5 (Phuong, Sach and Yen [15]) *Let X be a Banach space. We say that a locally Lipschitz function $\theta : K \subset X \to \mathbb{R}$ is **invex** on K, with respect to η if for any $x, y \in K$, there exists a vector $\eta(x, y) \in T_K(y)$ such that*

$$\theta(x) - \theta(y) - \theta^0(y; \eta(x, y)) \geq 0.$$

Definition 6 (Brandão, Rojas-Medar and Silva [2]) *Let X and Y be two Banach spaces and suppose that $P \subset Y$ is a convex cone. We say that $f : K \subset X \to Y$ is P-**invex** on K with respect to η, if $\omega^* \circ f : K \to \mathbb{R}$ is invex (in the sense of the previous definition) on K with respect to η, for each $\omega^* \in P^*$.*

Certainly, if $f : K \subset X \to \mathbb{R}$ is defifferentiable and convex, then f is invex and $\eta(x, y) = x - y$.

Note that when K is open, or more generally, if $y \in$ int K, we have $T_K(y) = X$, and, in this case, the above definition coincides with that given by Weir and Jeyakumar [21].

Moreover, when Y is infinite-dimensional and f is differentiable, the Definition 6 is the same to that given by Santos et al. in [17].

Recall that the set K *is called **invex** with respect to* η if there exists $\eta : K \times K \to X$ such that $y + \alpha \eta(x, y)$ is in K, for each $x, y \in K$ and $\alpha \in [0, 1]$.

In the sequence, we suppose that X and Y are two Banach spaces, K is a nonempty subset of X and $P \subset Y$ is a pointed convex cone such that $\operatorname{int} P \neq \emptyset$.

We consider the following vector variational-like inequality:

(VI) Find $x_0 \in K$ such that for each $x \in K$, there exists $\omega^* \in P^* \backslash \{0\}$ such that

$$(\omega^* \circ f)^0(x_0; \eta(x, x_0)) \geq 0.$$

Under suitable hypotheses, each solution of (VI) is a weakly efficient solution of (P1). In effect, we have:

Theorem 7 *Let K be an invex set with respect to η and $f : K \subset X \to Y$ be a P-invex function with respect to η. Then, each solution of (VI) is a weakly efficient solution of (P1).*

Proof. Suppose that x_0 is solution of (VI) which is not a weakly efficient solution of (P1). Then there exists $x \in K$ such that

$$f(x) - f(x_0) \in -\operatorname{int} P. \tag{3}$$

On the other hand, there exists $\omega^* \in P^* \backslash \{0\}$ such that $(\omega^* \circ f)^0(x_0; \eta(x, x_0)) \geq 0$. Since f is P-invex, it follows from (3) that

$$\omega^* \circ f(x) - \omega^* \circ f(x_0) \geq (\omega^* \circ f)^0(x_0; \eta(x, x_0)) \geq 0. \tag{4}$$

On the other hand, it follows from (3) that

$$\langle \omega^*, f(x) - f(x_0) \rangle = \omega^* \circ f(x) - \omega^* \circ f(x_0) < 0$$

and it contradicts (4). ∎

The following Lemma will be useful in the proof of our main result of this section.

Lemma 8 *[KKM-Fan Theorem [7]] Let X be a topological vector space, $E \subset X$ be a nonempty set and $F : E \rightrightarrows X$ is a multifunction such that for each $x \in E$, the set $F(x)$ is closed and nonempty. Moreover, suppose that there exists some $x \in E$ such that $F(x)$ is compact. If for each finite subset $\{x_1, ..., x_n\}$ of E one has*

$$\operatorname{co}\{x_1, ..., x_n\} \subset \bigcup_{i=1}^{n} F(x_i)$$

(where $\operatorname{co}\{x_1, ..., x_n\}$ is the convex hull of $\{x_1, .., x_n\}$) then

$$\bigcap_{x \in E} F(x) \neq \emptyset.$$

Now, we establish our existence result for (P1).

Theorem 9 *Let X be a reflexive Banach space and K a closed and bounded subset of X. Let $f : K \to Y$ be a strongly compactly Lipschitz function, P-invex with respect to η. Suppose that for each $x \in K$ and each $\omega^* \in P^*\backslash\{0\}$, the sets*

$$\Phi(x, \omega^*) := \{y \in K : (\omega^* \circ f)^0(x; \eta(y, x)) < 0\}$$

are convex. Furthermore, assume that η is continuous and $\eta(x, x) = 0$ for each $x \in K$. Then, the problem (P1) has a weakly efficient solution.

Proof. For $y \in K$ and $\omega^* \in P^*\backslash\{0\}$ we define

$$F(y, \omega^*) := \{x \in K : (\omega \circ f)^0(x; \eta(y, x)) \geq 0\}.$$

By Theorem 7, it is sufficient to prove that the variational-like inequality (VI) has a solution, that is,

$$\bigcap_{y \in K} \bigcup_{\omega^* \in P^*\backslash\{0\}} F(y, \omega^*) \neq \emptyset.$$

To do this, we will prove that

$$\bigcap_{y \in K} \bigcap_{\omega^* \in P^*\backslash\{0\}} F(y, \omega^*) \neq \emptyset.$$

Consider the multifunction $G : K \rightrightarrows X$ defined by

$$G(y) = \bigcap_{\omega^* \in P^*\backslash\{0\}} F(y, \omega^*).$$

We will prove that $\bigcap_{y \in K} G(y) \neq \emptyset$ by using of Lemma 8.

Suppose that the space X is equipped with the weak topology. Note that $G(y)$ is nonempty for all $y \in K$. In fact, $\eta(y, y) = 0$ and thus $y \in F(y, \omega^*)$ for all $\omega^* \in P^*\backslash\{0\}$. Hence, $y \in G(y)$. Furthermore, $G(y)$ is closed. In fact, let $y \in K$ be given and $(x_k) \subset G(y)$ is a sequence such that $x_k \to x$. Then, $(\omega^* \circ f)^0(x_k; \eta(y, x_k)) \geq 0$ for all $\omega^* \in P^*\backslash\{0\}$. Fixed $\omega^* \in P^*\backslash\{0\}$ we have

$$\limsup_{k \to \infty} (\omega^* \circ f)^0(x_k; \eta(y, x_k)) \geq 0. \tag{5}$$

But the function $(\omega^* \circ f)^0(\cdot, \cdot)$ is upper semicontinuous (see Proposition 1). Hence, it follows from (5) that

$$(\omega^* \circ f)^0(x; \eta(y, x)) \geq 0$$

that is, $x \in G(y)$ and thus $G(y)$ is closed. But $G(y)$ is convex and hence, weakly closed.

On the other hand, K is weakly compact because it is a convex, closed and bounded subset of X and X is a reflexive space. Then, $G(y)$ is weakly compact.

Now, take $x_1, ..., x_n \in K$. We should prove that $\text{co}\{x_1, ..., x_n\} \subset \bigcup_{i=1}^{n} G(x_i)$. Suppose that it is false. Then there exist $\alpha_i \geq 0$, $i = 1, ..., n$ such that $\sum_{i=1}^{n} \alpha_i = 1$ and $x := \sum_{i=1}^{n} \alpha_i x_i \notin \bigcup_{i=1}^{n} G(x_i)$. Hence, for each i, there exists $\omega_i^* \in P^* \backslash \{0\}$ such that $x \notin F(x_i, \omega_i^*)$, that is, $(\omega_i^* \circ f)^0(x; \eta(x_i, x)) < 0$. Next, we construct a $\omega^* \in P^* \backslash \{0\}$ such that $x_i \in \Phi(x, \omega^*)$ for all i, which will contradict the convexity of $\Phi(x, \omega^*)$, since $\eta(x, x) = 0$.

Define for $i, j \in \{1, ..., n\}, m \in \mathbb{N}$,

$$\omega_{i,j}^{(m)} = \begin{cases} \omega_j^* & \text{if } (\omega_j \circ f)^0(x; \eta(x_i, x)) < 0 \\ \frac{1}{m}\omega_j^* & \text{if } (\omega_j \circ f)^0(x; \eta(x_i, x)) \geq 0 \end{cases}$$

$$\omega_i^{(m)} = \sum_{j=1}^{n} \omega_{i,j}^{(m)}.$$

Clearly $\omega_i^{(m)}$ are linear functionals and for each $m, \omega_i^{(m)}$ are not all zero, for $i = 1, ..., n$. Furthermore, $\omega_i^{(m)}$ are continuous. In fact, define:

$$J_1 = \{j : (\omega_j^* \circ f)^0(x; \eta(x_i, x) < 0\}, \alpha_1 = \sharp J_1;$$
$$J_2 = \{j : (\omega_j^* \circ f)^0(x; \eta(x_i, x) \geq 0\}, \alpha_2 = \sharp J_2;$$
$$M = \max_{j=1,...,n} ||\omega_j^*||.$$

where $\sharp J_k$ denotes the cardinality of the set J_k.

Then, for all $u \in Y$, we have

$$|\omega_i^{(m)}(u)| = |\sum_{j \in J_1} \omega_j^*(u) + \frac{1}{m}\sum_{j \in J_j} \omega_j^*(u)|$$

$$\leq \sum_{j \in J_1} ||\omega_j^*||||u|| + \frac{1}{m}\sum_{j \in J_2} ||\omega_j^*||||u|| = (\alpha_1 + \frac{1}{m}\alpha_2)M||u||$$

that is, $\omega_i^{(m)} \in Y^*$. Furthermore, $\omega_i^{(m)}(u) \geq 0$, for each $u \in P$ (because $\omega_i^{(m)}$ is a nonnegative linear combination of functionals in P^*).

On the other hand,

$$(\omega_i^{(m)} \circ f)^0(x; \eta(x_i, x)) = [\sum_{j \in J_1} \omega_j^* \circ f + \frac{1}{m}\omega_j^* \circ f]^0(x, \eta(x_i, x))$$

$$\leq \sum_{j \in J_1}(\omega_j^* \circ f)^0(x; \eta(x_i, x)) + \frac{1}{m}\sum_{j \in J_2}(\omega_j \circ f)^0(x; \eta(x_i, x))$$

and hence, by taking m sufficiently big, say, $m \geq m(i)$, we have

$$(\omega_i^{(m)} \circ f)^0(x; \eta(x_i, x)) \leq \sum_{j \in J_1} (\omega_j^* \circ f)^0(x; \eta(x_i, x)) < 0, \ \forall m \geq m(i).$$

Take $M = \max_{i=1,...,n} m(i)$ and define $\omega^{(M)} = \sum_{i=1}^n \omega_i^{(M)}$. Then, we have that $\omega^{(M)} \in P^* \setminus \{0\}$ and

$$(\omega^{(M)} \circ f)^0(x; \eta(x_i, x)) \leq \sum_{i=1}^n (\omega_i^{(M)} \circ f)^0(x; \eta(x_i, x)) < 0$$

and then $x_i \in \Phi(x, \omega^{(M)})$ for all $i = 1, .., n$. But $\Phi(x, \omega^{(M)})$ is convex and thus $x \in \Phi(x, \omega^{(M)})$ but it contradicts $\eta(x, x) = 0$. Therefore, it follows from Lemma 8 that $\bigcap_{y \in K} G(y) \neq \emptyset$ and the proof is complete. $\blacksquare$

Remark 10 *We note that if K is compact and f is continuous, then (P1) has a weakly efficient solution (see Sawaragi et al. [19]). Thus, if we apply the Theorem 9 to the problem (P2), we do not obtain any new result, because the hypotheses considered here imply that K is compact and f is continuous (see Brézis [3]).*

4   The finite-dimensional problem (P2)

In this Section, we will consider a variational-like inequality of (weak) Minty type and we establish a characterization of weak efficiency for (P2) in terms of the solutions of this inequality. Using this result and a fixed point theorem for multifunctions, we establish our existence result for (P2).

We will consider the following vector variational-like inequality:

(WMVLI) Find $y \in X$ such that, for each $x \in X$ and $\xi_i \in \partial f_i(x)$, $i = 1, ..., p$

$$(\xi_1^T \eta(x, y), ..., \xi_p^T \eta(x, y)) \notin -\operatorname{int} \mathbb{R}_+^p.$$

In [16] Santos et al. proved that, under certain hypotheses, the solutions of (WMVLI) are weakly efficient solutions of (P2). In effect:

Proposition 11 *[Santos et al. [16]] Let X be a nonempty set of $\mathbb{R}^n$, invex with respect to η and $f_i : \mathbb{R}^n \to \mathbb{R}$, $i = 1, ..., p$ are locally Lipschitz functions and invex with respect to η. Suppose that η is skew (that is, $\eta(x, y) = -\eta(y, x), \forall x, y \in X$). Then, $y \in X$ is a weakly efficient solution of (P2) if and only if, $y \in X$ is a solution of (WMVLI).*

Now, we recall a fixed-point theorem of Fan-Browder, whose proof can be found in Park [14].

Lemma 12 *[Park [14]] Let X be a nonempty convex subset of a Hausdorff topological vector space E and let K be a nonempty compact subset of X. Suppose that $A, B : X \rightrightarrows X$ are multifunctions satisfying the following conditions:*

1. *$A(x) \subset B(x)$ for all $x \in X$;*

2. *$B(x)$ is a convex set for all $x \in X$;*

3. *$A(x) \neq \emptyset$, for all $x \in K$;*

4. *$A^{-1}(y) = \{x \in X : y \in A(x)\}$ is an open set for each $y \in X$;*

5. *for each finite subset N of X, there exists a compact, convex and nonempty subset L_N such that $L_N \supset N$ and $A(x) \cap L_N \neq \emptyset$ for all $x \in L_N \backslash K$.*

Then, there is a point $\overline{x} \in B(\overline{x})$ (that is, $\overline{x}$ is a fixed point of B).

Next, we will use Proposition 11 and Lemma 12 for to establish our result about the existence of weakly efficient solution for (P2), under a weaker compactness hypothesis on the feasible set X.

Theorem 13 *Let X be a nonempty and invex subset of $\mathbb{R}^n$ with respect to η and $f_i : \mathbb{R}^n \to \mathbb{R}$, $i = 1, .., p$ are locally Lipschitz functions and invex with respect to η. Assume that η is skew and such that $\eta(\cdot, y)$ is convex and continuous, for each $y \in X$. Moreover, suppose that for each finite subset N of $\mathbb{R}^n$, there exists a compact, convex and nonempty set $L_N \subset X$ such that $L_N \supset N$ and for all $x \in L_N \backslash K$, there is $z \in L_N$ such that there exist $\xi_i \in \partial f_i(z)$, $i = 1, ..., p$ satisfying*

$$(\xi_1^T \eta(z, x), ..., \xi_p^T \eta(z, x) \in -\operatorname{int} \mathbb{R}_+^p.$$

Then, problem (P2) has a weakly efficient solution.

Proof. For sake of readability, we give a proof using a concise notation. We denote by $\partial f(x)$ the set $\partial f_1(x) \times ... \times \partial f_p(x)$, $x \in X$. Let $s = (s_1, ..., s_p)$, where $s_i \in \mathbb{R}^n$, $i = 1, ..., p$. We denote by $s^T \eta(x, y)$ the vector

$$(s_1^T \eta(x, y), ..., s_p^T \eta(x, y)) \in \mathbb{R}^p.$$

Let $A, B : X \rightrightarrows X$ be multifunctions defined by:

$$A(x) \quad : \quad = \{z \in X : \exists t \in \partial f(z), t^T \eta(z, x) \in -\operatorname{int} \mathbb{R}_+^p\};$$
$$B(x) \quad : \quad = \{x \in X : \forall t \in \partial f(x), t^T \eta(z, x) \in -\operatorname{int} \mathbb{R}_+^p\}.$$

By using of Lemma 12, we will prove that there exists $y \in K$ such that $A(y) = \emptyset$, that is, y is solution of (WMVLI) and then, by Proposition 11, we conclude the desired result.

We will prove that the multifunctions A and B satisfy the conditions (1),(2),(4) and (5) of Lemma 12 but B does not have a fixed point.

We show that the condition (1) of Lemma 12 holds. Take $x \in X$ and $z \in A(x)$. Then, exists $t = (\xi_1, ..., \xi_p) \in \partial f(z)$ such that

$$(\xi_1^T \eta(z, x), ..., \xi_p^T \eta(z, x)) \in -\operatorname{int} \mathbb{R}_+^p. \tag{6}$$

Let $s = (\widehat{\xi}_1, ..., \widehat{\xi}_p) \in \partial f(x)$. By using of the invexity of functions f_i and the skewness of η, we conclude that, for each $i = 1, ..., p$

$$
\begin{aligned}
\widehat{\xi}_i^T \eta(z, x) &\leq f_i^0(x; \eta(z, x)) \leq f_i(z) - f_i(x) = -(f_i(x) - f_i(z)) \\
&\leq -\widehat{\xi}_i^T \eta(x, z) = \widehat{\xi}_i^T \eta(z, x).
\end{aligned} \tag{7}
$$

Follows from (6) and (7),

$$(\widehat{\xi}_1^T \eta(z, x), ..., \widehat{\xi}_p^T \eta(z, x)) \in -\operatorname{int} \mathbb{R}_+^p,$$

and then $z \in B(x)$.

Now, we prove the second condition of Lemma 12: Let $x \in X$ be given and take $z_1, z_2 \in B(x)$ and $\lambda \in [0, 1]$. Then, for each $s = (\xi_1, ..., \xi_p) \in \partial f(x)$, we have

$$(\xi_1^T \eta(z_1, x), ..., \xi_p^T \eta(z_1, x)), \ (\xi_1^T \eta(z_2, x), ..., \xi_p^T \eta(z_2, x)) \in -\operatorname{int} \mathbb{R}_+^p. \tag{8}$$

For each $j = 1, ..., p$ we consider $\xi_j = (\xi_j^{(1)}, ..., \xi_j^{(n)})$, where $\xi_j^{(k)} \in \mathbb{R}$, $\eta(x, y) = (\eta_1(x, y), ..., \eta_n(x, y))$, $\eta_k(x, y) \in \mathbb{R}$. From the convexity of η_k and (8) we obtain

$$
\begin{aligned}
\xi_j^T \eta(\lambda z_1 + (1 - \lambda)z_2, x) &= \sum_{k=1}^{n} \xi_j^{(k)} \eta_k(\lambda z_1 + (1 - \lambda)z_2, x) \\
&\leq \sum_{k=1}^{n} \xi_j^{(k)} [\lambda \eta_k(z_1, x) + (1 - \lambda)\eta_k(z_2, x)] \\
&= \lambda \xi_j^T \eta(z_1, x) + (1 - \lambda)\xi_j^T \eta(z_2, x) < 0
\end{aligned}
$$

for each $j = 1, .., p$. Hence, $\lambda z_1 + (1 - \lambda)z_1 \in B(x)$.

The fourth condition is proved as follows: For all $z \in X$, we will show that the set $(A^{-1}(z))^c$ is closed. To do this, consider a sequence $(x_n) \subset (A^{-1}(z))^c$ such that x_n converges to x. Then $x_n \notin A^{-1}z$, for all $n \in N$. Take $t = (\xi_1, ..., \xi_p) \in \partial f(z)$ such that

$$(\xi_1^T \eta(z, x_n), ..., \xi_p^T \eta(z, x_n)) \notin -\operatorname{int} \mathbb{R}_+^p. \tag{9}$$

Because $\eta(\cdot, z)$ is continuous and skew, we have that $\eta(z, \cdot)$ is continuous too and since $(-\operatorname{int} \mathbb{R}_+^p)^c$ is a closed set, by taking $n \to \infty$ in (9) we obtain $(\xi_1^T \eta(z, x), ..., \xi_p^T(z, x)) \notin -\operatorname{int} \mathbb{R}_+^p$ and thus $x \in (A^{-1}(z))^c$.

From the hypotheses done, the condition (5) of Lemma 12 also holds.

However, B does not have a fixed point, because otherwise it would exists some $x \in X$ such that for all $s \in \partial f(x)$, and $s^T \eta(x,x) = 0 \in - \operatorname{int} \mathbb{R}^p_+$, which is absurd. Thus, from Lemma 12, it follows that there exists $y \in K$ such that $A(y) = \emptyset$. ∎

We have the following consequence of Theorem 13:

Corollary 14 *Let X be a nonempty subset of $\mathbb{R}^n$, invex with respect to η and suppose that η is skew and such that $\eta(\cdot, y)$ is convex and continuous. Moreover, assume that*

$$K = \{x \in X : (f_1^0(z_0; \eta(z_0, x)), ..., f_p^0(z_0; \eta(z_0, x))) \notin - \operatorname{int} \mathbb{R}^p_+\}$$

is compact for some $z_0 \in X$. Then, (P2) has a weakly efficient solution.

Proof. Let N be a nonempty and finite subset of X. Define:

$$L_N := \overline{\operatorname{co}}(N \cup K)$$

(where $\overline{\operatorname{co}}$ denotes the closed convex hull). Then, for each $x \in L_N \backslash K$, we have

$$(f_1^0(z_0; \eta(z_0, x)), ..., f_p^0(z_0; \eta(z_0, x))) \in - \operatorname{int} \mathbb{R}^p_+. \tag{10}$$

Take $z = z_0 \in K \subset L_N$ and $\xi_i \in \partial f_i(z)$. We have,

$$\xi_i^T \eta(z_0, x) \le f_i^0(z_0; \eta(z_0, x)), \ i = 1, .., p \tag{11}$$

and from (10) and (11), we obtain

$$(\xi_1^T \eta(z, x), ..., \xi_p^T \eta(z, x)) \in - \operatorname{int} \mathbb{R}^p_+.$$

Thus, by Theorem 13, (P2) has a weakly efficient solution. ∎

5   Conclusions

In this note, we obtain an existence theorem for weakly efficient solutions of vector optimization problems defined between infinite-dimensional Banach spaces whose objective function is invex and strongly compactly Lipschtz. We characterize the solutions of this problem in terms of the solutions of a variational-like inequality and, by applying this characterization an d the KKM-Fan Theorem, we establish our result. The approach that we used here is very similar with to the one employed by Chen and Craven [4] where the authors considered the differentiable problem. Also, we consider the vector problem defined between finite-dimensional spaces, and, for this problem we also obtain a result on existence of weakly efficient solution. Our approach is analogous to the one used by Lee [13], where it was considered the non differentiable convex case. We use a characterization of the weakly efficiency in terms of the solutions of the (weak) Minty type inequalities and a fixed point theorem for multifunctions for to prove our result. The results that we present here generalize those obtained by Chen and Craven [4] and Lee [13].

References

[1] C. Baiocchi and A. Capelo, Variational and quasivariational inequalities: applications to the free boundary problems, John Wiley and Sons Inc., New York, 1984.

[2] A.J.V. Brandão, M.A. Rojas-Medar and G.N. Silva, Optimality conditions for Pareto nonsmooth nonconvex programming in Banach spaces, J. Optim. Th. Appl. 103 (1) (1999) 65-73.

[3] H. Brézis, Analyse Fonctionelle: Théorie et applications, Masson, Paris, 1983.

[4] G.Y. Chen and B.D. Craven, Existence and continuity of solutions for vector optimization, J. Optim.Th. Appl. 81 (1984) 459-468.

[5] F.H. Clarke, Optimization and Nonsmooth Analysis, Willey -Interscience, 1983.

[6] G. Duvault and J.L. Lions, Les inéquations en Mecanique et en Physique, Dunod, Paris, 1972.

[7] K. Fan, A generalization of Tichonoff´s fixed point theorem, Mathematische Annalen 142 (1961) 305-310.

[8] F. Giannessi, Theorems of the alternative and complementarity problems (Proceedings of the International School, Erice, 1978), John Willey, Chichester, England, (1980) 151-186.

[9] F. Giannessi, On Minty variational principle, in: New trends in Mathematical programming, Kluwer, (1997) 93-99.

[10] I.V. Girsanov, Lectures on Mathematical Theory of Extremum Problems. Lecture Notes in Economics and Mathematical Systems, vol. 67, Springer-Verlag, Berlin, Germany, 1972.

[11] R. Glowinsky, J.L. Lions and R. Trémoliéres, Numerical Analysis of Variational Inequalities, North-Holland, 1976.

[12] P.T. Parker and J.S. Pang, Finite-dimensional variational inequality and nonlinear complementarity problems: a survey of theory, algorithms and applications, Math. Program. 48 (1990) 161-220.

[13] G.M. Lee, On relations between vector variational inequality and vector optimization problem, in: X. Yang, A.I. Mees, M. Fisher and L. Jennings (Eds.), Progress in Optimization, Kluwer, 2000.

[14] S. Park, Some coincidence theorems on acyclic multifunctions and applications to KKM Theory, in: Fixed Point Theory and Applications, World Scientific Publishing, River Edge, NJ, (1992) 248-277.

[15] T.D. Phuong, P.H. Sach and N.D. Yen, Strict lower semicontinuity of the level sets and invexity of a locally Lipschitz function, J. Optim. Th. Appl. 87 (1995) 579-594.

[16] L.B. Santos, M. Rojas-Medar and A. Rufián-Lizana, Some relations between variational-like inequalities and efficient solutions of certain nonsmooth optimization problems, Int. J. Math. Sci. 2006 (2006), 16 (article ID 26808), doi: 101155/IJMMS/2006/26808.

[17] L.B. Santos, R. Osuna-Gómez, M.A. Rojas-Medar and A. Rufián-Lizana, Preinvex functions and weak efficient solutions for some vectorial optimization problems in Banach spaces, Comput. Math. Appl. 48 (2004) 885-895.

[18] L.B. Santos, M. Rojas-Medar, G. Ruiz-Garzón and A. Rufián-Lizana, Existence of weakly efficient solutions in nonsmooth vector optimization, Appl. Math. Comput. 200 (2008) 547-556.

[19] Y. Sawaragi, H. Nakayama and T. Tanino, Theory of Multiobjective Optimization, Academic Press, New York, 1985.

[20] L. Thibault, On generalized differentials and subdifferentials of Lipschitz vector-valued functions, Nonlinear Anal. 6 (1980) 1037-1053.

[21] T. Weir and V. Jeyakumar, A class of nonconvex functions and mathematical programming, Bull. Austral. Math. Soc. 38 (1988) 177-189.

CHAPTER 5

Proper efficiency and duality for differentiable multiobjective programming problems with B-(p, r)-invex functions

Tadeusz Antczak*

Abstract

In this paper, the concept of proper efficiency has been incorporated to develop the optimality conditions and duality results for differentiable nonconvex multiobjective programming problems. The assumptions on properly efficiency are relaxed by means of the introduced classes of vector-valued B-(p, r)-invex functions. These results extend several known results to a wider class of nonconvex vector optimization problems.

Keywords: multiobjective programming, properly effcient solution, (strictly) B-(p, r)-invex function with respect to η and b, optimality conditions, duality.

1 Introduction

In recent years, much attention was paid to the theory of multiobjective optimization, because the models with multiple objectives can better describe many complicated economics problems. A vector minimum problem arises when two or more functions are to be mini mized over a feasible region. Generally, the concept of an optimal solution valid in scalar mathematical programming problems does not work in vector optimization problems. The more general concept of efficiency or Pareto optimality in multiobjective programming has important role in all optimal decision problems with noncomparable criteria. However, to eliminate the undesirable anomaly of efficient solutions, Kuhn and Tucker [19] introduced the concept of a noninferior proper solution.

Geoffrion [12] reformulated this concept slightly and defined the proper efficient solutions for a multiobjective programming problem. He introduced this concept for the purpose of eliminating an undesirable possibility in the concept of efficiency, namely the possibility of the criterion functions being such that efficient solutions could be found for which the marginal gain for one function could be made arbitrarily large relative to the marginal losses for the others.

Using this concept, a number of researches including Das and Nanda [10], Egudo and Hanson [11], Henig [16], Taninio and Sawaragi [23], Weir [24], Weir and Mond [25], White

---

*Faculty of Mathematics and Computer Science, University of Lodz, Banacha 22, 90-238 Lodz, Poland. e-mail: antczak@math.uni.lodz.pl

[26] and others have developed proper efficiency optimality conditions and duality concepts in multiobjective programming.

But in most of such studies an assumption of convexity on the problems was made. Recently, several classes of functions have been defined for the purpose of weakening the limitations of convexity in mathematical programming. One of the useful generalizations is the concept of invexity which was originated by Hanson [13], but so named by Craven [8].

Hanson's initial results inspired a great deal of subsequent work which has greatly expanded the role of invexity in optimization. Ben-Israel and Mond [7], Craven and Glover [9], Hanson and Mond [14], Bector and et al. [6], Jeyakumar and Mond [17], Martin [20] and many others have studied some properties, applications and further generalizations of these functions. One of such generalizations of invex functions is (p, r)-invexity, which was introduced by Antczak [1] in the scalar case and was extended to the vectorial case in [3]. Later, Antczak [2] generalized the concept of (p, r)-invexity and the concept of B-invexity [6] and he introduced the class of scalar B-(p, r)-invex functions. He gave some necessary and sufficient conditions for B-(p, r)-invexity and showed the relationships between the defined class of B-(p, r)-invex functions with respect to η and other classes of (generalized) invex functions. Further, Antczak obtained some sufficient optimality conditions and duality results for scalar optimization problems with B-(p, r)-invex functions.

In the present paper, we extend, in natural way, a class of B-(p, r)-invex functions, previously defined by Antczak [2] for scalar mathematical programming problems, to the vectorial case. In this way, we define a new class of differentiable nonconvex vector valued functions, namely vector-valued B-(p, r)-invex vector functions. However, the main purpose of this article is to apply B-(p, r)-invexity to develop optimality conditions and duality theory for differentiable multiobjective programming problems. Considering the concept of a properly efficient solution [12] in multiobjective programming problems, we establish sufficient optimality conditions for vector optimization problems in which the involved functions belong to the class of B-(p, r)-invex functions (with respect to the same function η and with respect to, not necessarily, the same function b). Moreover, duality results in Mond-Weir format are established by using a properly efficient type relation between the primal and dual objective functions and the concept of vectorial B-(p, r)-invexity. Furthermore, we consider two types of dual problems in Wolfe format. One of them is a so-called Wolfe dual problem with vector constraints and the second one is called a Wolfe dual problem with scalar constraints. In the case of these nonconvex vector dual problems, we prove various duality results under vectorial B-(p, r)-invexity imposed on the modified vector-valued Lagrange function defined for the considered multiobjective programming problem. Moreover, various duality theorems are established for Wolfe dual problem with scalar constraints also under B-(p, r)-invexity assumption of the scalar Lagrange function and, what is interesting, under suitable B-(p, r)-invexity assumptions of the functions constituting the considered multiobjective programming problem. The optimality results obtained in the paper are illustrated by a suitable example of a multiobjective optimization problem with vector B-(p, r)-invex functions with respect to the same function η and with respect to, not necessarily, the same function b.

2   New classes of nonconvex vector-valued functions

The following convention for equalities and inequalities will be used throughout the paper.

For any $x = (x_1, x_2, ..., x_n)$, $y = (y_1, y_2, ..., y_n)$, where $n > 1$, we define:

(i) $x = y$ if and only if $x_i = y_i$ for all $i = 1, 2, ..., n$;

(ii) $x < y$ if and only if $x_i < y_i$ for all $i = 1, 2, ..., n$;

(iii) $x \leqq y$ if and only if $x_i \leqq y_i$ for all $i = 1, 2, ..., n$;

(iv) $x \leq y$ if and only if $x \leqq y$ and $x \neq y$.

Throughout the paper, we will use the same notation for row and column vectors when the interpretation is obvious.

Let $\lambda \in R^k$. Then $diag\, \lambda$ denotes the following diagonal matrix of $k \times k$, that is,

$$
diag\, \lambda = \begin{bmatrix} \lambda_1 & 0 & \cdots & 0 \\ 0 & \ddots & & \vdots \\ \vdots & & \ddots & 0 \\ 0 & \cdots & 0 & \lambda_k \end{bmatrix}.
$$

The following definition generalizes the definition of a class of scalar B-(p,r)-invex functions [2] to the case of a class of vector-valued B-(p,r)-invex functions.

Definition 1 *Let X be a nonempty open subset of R^n. The differentiable vector-valued function $f : X \to R^k$ is said to be vector B-(p,r)-invex (strictly B-(p,r)-invex) with respect to η at $u \in X$ on X if, there exist real numbers p, r, a function $\eta : X \times X \to R^n$ and a function $b : X \times X \to R^k_+$ such that, for all $x \in X$, the inequality*

$$
\begin{aligned}
\tfrac{1}{r}diag\, b(x,u) \left(e^{r(f(x)-f(u))} - 1 \right) &\geqq \tfrac{1}{p}\nabla f(u) \left(e^{p\eta(x,u)} - 1 \right) & (> \text{ if } x \neq u) \quad p \neq 0, r \neq 0, \\
\tfrac{1}{r}diag\, b(x,u) \left(e^{r(f(x)-f(u))} - 1 \right) &\geqq \nabla f(u)\eta(x,u) & (> \text{ if } x \neq u) \quad p = 0, r \neq 0, \\
diag\, b(x,u) \left(f(x) - f(u) \right) &\geqq \tfrac{1}{p}\nabla f(u) \left(e^{p\eta(x,u)} - 1 \right) & (> \text{ if } x \neq u) \quad p \neq 0, r = 0, \\
diag\, b(x,u) \left(f(x) - f(u) \right) &\geqq \nabla f(u)\eta(x,u) & (> \text{ if } x \neq u) \quad p = 0, r = 0,
\end{aligned}
\tag{1}
$$

holds.

If the inequality (1) is satisfied at each point $u \in X$, then f is said to be vector B-(p,r)-invex (strictly B-(p,r)-invex) with respect to η on X.

It should be pointed out that the exponentials appearing in inequalities (1) are understood to be taken componentwise and $\mathbf{1} = (1, 1, ..., 1) \in R^n$.

Remark 1 *In order to define an analogous class of (strictly) B-(p,r)-incave vector functions with respect to η, the direction of the inequalities in (1) should be changed to the opposite one.*

Now, we write down the above inequalities (1) defining the various classes of b_i-(p, r)-invex functions with respect to each component of the vector-valued function f of this type.

Definition 2 *Let $f : X \to R^k$ be a differentiable vector-valued function defined on a nonempty open set $X \subset R^n$. If, there exist real numbers p, r, a function $\eta : X \times X \to R^n$ and functions $b_i : X \times X \to R_+$, $i = 1, ..., k$, such that for all $x \in X$ and any $i = 1, ..., k$,*

$$\frac{1}{r} b_i(x, u) \left(e^{r(f_i(x) - f_i(u))} - 1 \right) \geqq \frac{1}{p} \nabla f_i(u) \left(e^{p\eta(x,u)} - \mathbf{1} \right) \; (> \text{ if } x \neq u) \text{ for } p \neq 0, r \neq 0$$

$$\frac{1}{r} b_i(x, u) \left(e^{r(f_i(x) - f_i(u))} - 1 \right) \geqq \nabla f_i(u)\eta(x, u) \qquad (> \text{ if } x \neq u) \text{ for } p = 0, r \neq 0$$

$$b_i(x, u) \left(f_i(x) - f_i(u) \right) \geqq \frac{1}{p} \nabla f_i(u) \left(e^{p\eta(x,u)} - \mathbf{1} \right) \quad (> \text{ if } x \neq u) \text{ for } p \neq 0, r = 0 \qquad (2)$$

$$b_i(x, u) \left(f_i(x) - f_i(u) \right) \geqq \nabla f_i(u)\eta(x, u) \qquad (> \text{ if } x \neq u) \text{ for } p = 0, r = 0$$

then f_i, $i = 1, ..., k$, is said to be a b_i-(p, r)-invex vector function (a strictly vector b_i-(p, r)-invex function) with respect to η at u on X.

If the inequality (2) is satisfied at each point $u \in X$, then f_i, $i = 1, ..., k$, is said to be b_i-(p, r)-invex (strictly b_i-(p, r)-invex) with respect to η on X.

Remark 2 *The introduced definition of a vector-valued B-(p, r)-invex function is an extension of various well-known classes of nonconvex functions:*

   i) For $k = 1$, Definition 1 reduces to the definition of a scalar B-(p, r)-invex function (see [2]).

   ii) In the case $b(x, u) \equiv \mathbf{1}$, we obtain the definition of a vector-valued (p, r)-invex function (see [3]).

  iii) If $p = 0$ and $r = 0$, we get the definition of a vector-valued B-invex function introduced by Bector et al. (see [6])

  iv) For $k = 1$, $b(x, u) \equiv \mathbf{1}$, $p = 0$ and $r = 0$, Definition 1 reduces to the definition of a scalar invex function introduced by Hanson (see [13]).

   v) In the case when $\eta(x, u) = x - u$, we obtain a definition of a vector-valued B-(p, r)-convex function.

  vi) If $\eta(x, u) = x - u$ and $b(x, u) \equiv \mathbf{1}$, we get the definition of a vector-valued (p, r)-convex function.

 vii) For $k = 1$, $b(x, u) \equiv \mathbf{1}$, $p = 0$, we obtain the definition of a scalar r-invex function introduced by Antczak [4].

viii) For $k = 1$, $\eta(x, u) = x - u$, $b(x, u) \equiv \mathbf{1}$, $p = 0$, Definition 1 reduces to the definition of a scalar r-convex function introduced by Avriel (see [5]).

Remark 3 *All theorems in the further part of this work will be established only in the case when $p \neq 0$, $r \neq 0$ (other cases can be dealt with likewise since the only changes arise from the form of inequality defining the class of B-(p, r)-invex functions with respect to η for given p and r). The proofs in the other cases are easier than those in the considered one. This follows from the form of inequalities which are given in Definition 1. Moreover, without loss of generality, we shall assume that $r > 0$ (in the case when $r < 0$, the direction some of the inequalities in the proofs of the given theorems should be changed to the opposite one).*

Proposition 1 *Let $f : R^n \rightarrow R^k$ be a vector B-(p, r)-invex (B-(p, r)-incave) function with respect to η on R^n, where $\eta : R^n \times R^n \rightarrow R^n$ and $b = (b_1, ..., b_k) : R^n \times R^n \rightarrow R^k_+$. The following propositions are true:*
a) If α be any real number, then the function $f + \alpha$ is B-(p, r)-invex (B-(p, r)-incave) with respect to the same function η on R^n.
b) If λ is any positive real number, then λf_i, $i = 1, ..., k$, is b_i-$(p, \frac{r}{\lambda})$-invex (b_i-$(p, \frac{r}{\lambda})$-incave) with respect to the same function η on R^n.
c) f is B-(p, r)-invex (B-(p, r)-incave) with respect to η on R^n if and only if $-f$ is B-$(p, -r)$-incave (B-$(p, -r)$-invex) with respect to η on R^n.

The classes of scalar B-(p, r)-invex functions have been characterized by Antczak in [2].

3   Optimality conditions in multiobjective programming

We consider a constrained multiobjective programming problem

$$V\text{-minimize } f(x) = (f_1(x), f_2(x), ..., f_k(x))$$
$$\text{(VP)} \qquad g_j(x) \leqq 0, \ j = 1, ..., m,$$
$$x \in X,$$

where $f_i : X \rightarrow R$, $i \in I = \{1, ..., k\}$, $g_j : X \rightarrow R$, $j \in J = \{1, ..., m\}$ are differentiable functions on a nonempty open set $X \subset R^n$.

Let $D = \{x \in X : g_j(x) \leqq 0, \ j \in J\}$ be the set of all feasible solutions for the considered multiobjective programming problem (VP) and $J(z) := \{j \in J : g_j(z) = 0\}$ denote the index set of constraints active at $z \in D$.

Further, we define the following modified vector-valued Lagrange function for the considered multiobjective programming problem

$$L_k\,(x, \lambda, \xi) := diag\,\lambda f(x) + \frac{1}{k}\xi g(x)\mathbf{e}, \qquad (3)$$

where $e = (1, ..., 1) \in R^k$, and, moreover, the scalar Lagrange function for problem (VP)

$$L\,(x, \lambda, \xi) := \lambda f(x) + \xi g(x). \qquad (4)$$

For such optimization problems minimization means in general obtaining (weak) Pareto optimal solutions in the following sense [22]:

Definition 3 *A feasible point $\overline{x} \in D$ is said to be an efficient solution (Pareto solution) for (VP) if and only if there exists no $x \in X$ such that*

$$f(x) \leqq f(\overline{x}).$$

It is known that some efficient solutions present an undesirable property with respect to the ratio between the marginal profit of an objective function and the loss of some other. The concept of proper efficiency given by Geoffrion [12] is a slightly restricted definition of efficiency which eliminates efficient points of a certain anomalous type.

Definition 4 *An efficient solution $\overline{x}$ for which there exists a scalar $M > 0$, such that, for each $i = 1, ..., k$, we have*

$$\frac{f_i(x) - f_i(\overline{x})}{f_j(\overline{x}) - f_j(x)} \leqq M \tag{5}$$

for some j such that $f_j(x) > f_j(\overline{x})$ and $f_i(x) < f_i(\overline{x})$ for $x \in D$, is called a properly efficient solution of (VP).

The quantity in the relation (5) may be interpreted as the marginal trade-off for the objective functions f_i and f_j between x and $\overline{x}$. It can be made arbitrarily large for suitable choice of x with $f_j(x) > f_j(\overline{x})$ and $f_i(x) < f_i(\overline{x})$. It may be advisable to drop $\overline{x}$ from the possible choice of "optimal" solution though $\overline{x}$ may be efficient.

An efficient point that is not properly efficient is said to be improperly efficient.

To prove the Karush-Kuhn-Tucker necessary optimality conditions for the considered multiobjective programming problem, we introduce the following constraint qualification:

Generalized Slater Constraint Qualification (GSCQ): We said that the constraint function g satisfies (GSCQ) at $\overline{x} \in D$, if there exists $\widetilde{x} \in D$ such that $g(\widetilde{x}) < 0$ and g_j, $j \in J(\overline{x})$, are b_j-invex at $\overline{x}$ on D with respect to the same function η.

It is well-known (see, for example, [12], [15], [18]) that the following first-order optimality conditions are necessary for a feasible solution $\overline{x}$ in (VP) to be a properly efficient solution in this vector optimization problem.

Theorem 2 *Let $\overline{x}$ be a properly efficient point in problem (VP). Moreover, assume that g satisfies the generalized Slater constraint qualification at $\overline{x}$. Then, there exist $\overline{\lambda} \in R^k$ and $\overline{\xi} \in R^m$ such that*

$$\overline{\lambda} \nabla f(\overline{x}) + \overline{\xi} \nabla g(\overline{x}) = 0, \tag{6}$$

$$\overline{\xi} g(\overline{x}) = 0, \tag{7}$$

$$\overline{\lambda} > 0, \quad \sum_{i=1}^{k} \overline{\lambda} = 1, \quad \overline{\xi} \geqq 0. \tag{8}$$

Now, we prove the sufficient optimality conditions for $\overline{x}$ to be a properly efficient point in the considered multiobjective programming problem (VP). To do this, we assume that the functions constituting problem (VP) are B-(p,r)-invex at $\overline{x}$ on D with respect to the same function η and with respect to some functions b, not necessarily equal to each other for various objective and constraint functions.

Theorem 3 *Let $\overline{x}$ be a feasible point for (VP) and there exist the Lagrange multipliers $\overline{\lambda} \in R^k$ and $\overline{\xi} \in R^m$ such that the Karush-Kuhn-Tucker necessary optimality conditions (6)-(8) be satisfied at $\overline{x}$. Further, assume that f_i, $i \in I$, are b_{f_i}-(p,r)-invex with respect to the same function η at $\overline{x}$ on D, $b_{f_i}(x, \overline{x}) > 0$, $i \in I$, for all $x \in D$, and g_j, $j \in J(\overline{x})$ are b_{g_j}-(p,r)-invex with respect to the same function η at $\overline{x}$ on D. Then $\overline{x}$ is a properly efficient solution in (VP).*

Proof.   We proceed by contradiction. Suppose that $\overline{x}$ is not an efficient solution in problem (VP). Then, there exists a feasible solution $\widetilde{x}$ in (VP) such that

$$f(\widetilde{x}) \leq f(\overline{x}). \tag{9}$$

By assumption, any f_i, $i \in I$, is b_{f_i}-(p,r)-invex with respect to the same function η at $\overline{x}$ on D. Since $\overline{\lambda}_i > 0$ for $i \in I$, then, by Proposition 1 b), $\overline{\lambda}_i f_i$ is b_{f_i}-$(p, \frac{r}{\overline{\lambda}_i})$-invex with respect to the same functions η at $\overline{x}$ on D. Hence, by Definition 2, we get, for any $i \in I$,

$$\frac{\overline{\lambda}_i}{r} b_{f_i}(\widetilde{x}, \overline{x}) \left(e^{\frac{r}{\overline{\lambda}_i} \overline{\lambda}_i (f_i(\widetilde{x}) - f_i(\overline{x}))} - 1 \right) \geqq \frac{1}{p} \overline{\lambda}_i \nabla f_i(\overline{x}) \left(e^{p\eta(\widetilde{x}, \overline{x})} - 1 \right). \tag{10}$$

Since $b_{f_i}(\widetilde{x}, \overline{x}) > 0$ for any $i \in I$, then, by (9) and (10), for any $i \in I$,

$$\frac{1}{p} \overline{\lambda}_i \nabla f_i(\overline{x}) \left(e^{p\eta(\widetilde{x}, \overline{x})} - 1 \right) \leqq 0, \tag{11}$$

and for at least one $i_0 \in I$

$$\frac{1}{p} \overline{\lambda}_{i_0} \nabla f_{i_0}(\overline{x}) \left(e^{p\eta(\widetilde{x}, \overline{x})} - 1 \right) < 0. \tag{12}$$

Thus, by (11) and (12),

$$\frac{1}{p} \sum_{i=1}^{k} \overline{\lambda} \nabla f(\overline{x}) \left(e^{p\eta(\widetilde{x}, \overline{x})} - 1 \right) < 0. \tag{13}$$

By the Karush-Kuhn-Tucker necessary optimality conditions (7) and (8), it follows that for any $j \in J$ and for all $x \in D$, we have

$$\overline{\xi}_j g_j(x) \leqq \overline{\xi}_j g_j(\overline{x}). \tag{14}$$

By assumption, any function g_j, $j \in J(\overline{x})$, is b_{g_j}-(p, r)-invex with respect to the same function η at $\overline{x}$ on D. Thus, by Lemma 1 b), any function $\overline{\xi}_j g_j$, $j \in J(\overline{x})$ is b_{g_j}-$(p, \frac{r}{\overline{\xi}_j})$-invex with respect to the same function η at $\overline{x}$ on D. Moreover, by Definition 2, there exist $b_{g_j} : D \times D \to R$, $j \in J(\overline{x})$, such that $b_{g_j}(x, \overline{x}) \geq 0$ for all $x \in D$. Thus, (14) gives

$$\frac{1}{r} b_{g_j}(x, \overline{x}) \left(e^{r\left(\overline{\xi}_j g_j(x) - \overline{\xi}_j g_j(\overline{x})\right)} - 1 \right) \leq 0 \tag{15}$$

Using Definition 2 together with (15), we get, for any $j \in J(\overline{x})$ and for all $x \in D$,

$$\frac{1}{p} \overline{\xi}_j \nabla g_j(\overline{x}) \left(e^{p\eta(x, \overline{x})} - 1 \right) \leq 0.$$

Taking into account the constraint functions g_j, $j \notin J(\overline{x})$, we obtain, for $x = \widetilde{x}$,

$$\frac{1}{p} \overline{\xi} \nabla g(\overline{x}) \left(e^{p\eta(\widetilde{x}, \overline{x})} - 1 \right) \leq 0. \tag{16}$$

Thus, by (13) and (16), we get the following inequality

$$\frac{1}{p} \left(\overline{\lambda} \nabla f(\overline{x}) + \overline{\xi} \nabla g(\overline{x}) \right) \left(e^{p\eta(\widetilde{x}, \overline{x})} - 1 \right) < 0,$$

which contradicts the Karush-Kuhn-Tucker necessary optimality condition (6). This means that $\overline{x}$ is an efficient solution in (VP).

It remains to prove that $\overline{x}$ is a properly efficient solution in (VP). We assume that $k \geq 2$. Next, let

$$M = (k - 1) \max_{i,j} \frac{\overline{\lambda}_j}{\overline{\lambda}_i}.$$

We proceed by contradiction. Suppose that $\overline{x}$ is not a properly efficient point in (VP). Then, by Definition 4, there exists $\widetilde{x} \in D$ and the objective functions f_i and f_j such that, for every scalar $M > 0$, and $f_i(\widetilde{x}) - f_i(\overline{x}) < 0$, the following inequality

$$f_i(\overline{x}) - f_i(\widetilde{x}) > M \left(f_j(\widetilde{x}) - f_j(\overline{x}) \right), \tag{17}$$

holds for each $j \neq i$ such that $f_j(\widetilde{x}) > f_j(\overline{x})$.

By (17), it follows that, for each $j \neq i$,

$$f_i(\overline{x}) - f_i(\widetilde{x}) > (k - 1) \max_{i,j} \frac{\overline{\lambda}_j}{\overline{\lambda}_i} \left(f_j(\widetilde{x}) - f_j(\overline{x}) \right) > (k - 1) \frac{\overline{\lambda}_j}{\overline{\lambda}_i} \left(f_j(\widetilde{x}) - f_j(\overline{x}) \right).$$

Thus,

$$\frac{\overline{\lambda}_i}{k - 1} \left(f_i(\overline{x}) - f_i(\widetilde{x}) \right) > \overline{\lambda}_j \left(f_j(\widetilde{x}) - f_j(\overline{x}) \right). \tag{18}$$

Summing (18) over $j \neq i$, we obtain

$$\overline{\lambda}_i \left(f_i \left(\overline{x} \right) - f_i \left(\widetilde{x} \right) \right) > \sum_{j \neq i} \overline{\lambda}_j \left(f_j \left(\widetilde{x} \right) - f_j \left(\overline{x} \right) \right).$$

Hence, the following inequality

$$\overline{\lambda}_i f_i \left(\overline{x} \right) + \sum_{j \neq i} \overline{\lambda}_j f_j \left(\overline{x} \right) > \overline{\lambda}_i f_i \left(\widetilde{x} \right) + \sum_{j \neq i} \overline{\lambda}_j f_j \left(\widetilde{x} \right)$$

holds, which is a contradiction to the efficiency of $\widetilde{x}$. Thus, the conclusion of this theorem is established. ∎

To illustrate the results established in the paper, we consider an example of a differentiable nonconvex multiobjective programming problem and we show that to establish the optimality of some feasible point in this vector optimization problem, the sufficient optimality conditions from Theorem 3 are applicable.

Example 1 *We now consider the following multiobjective programming problem*

$$V\text{-}minimize\ f(x) = (f_1(x),\ f_2(x)) = \left(\ln \left[\left(\arctan^2 x + 1 \right) \frac{e^x - 1}{2} + 1 \right],\ \ln \left(e^{x-1} + x \right) \right)$$

$$g(x) = \ln \left(e^{-x} \left(x^2 - x \right) + 1 \right) \leqq 0.$$

Note that the set of all feasible solutions is $D = \{x \in R : 0 \leqq x \leqq 1\}$ and the unique properly efficient solution point is $\overline{x} = 0$. To verify the optimality of the feasible solution $\overline{x} = 0$, we use the sufficient optimality conditions from Theorem 3. Note that the objective functions f_i, $i = 1, 2$, are strictly b_{f_i}-$(1, 1)$-invex with respect to the same function η defined by

$$\eta(x, \overline{x}) = \ln (x + 1) - \ln (\overline{x} + 1)$$

and with respect to the functions b_{f_1} and b_{f_2}, respectively, defined by:

$$b_{f_1}(x, \overline{x}) = \frac{1}{\arctan^2 x + 1}\ ,\quad b_{f_2}(x, \overline{x}) = 1 + e$$

Also the constraint function g is a b_g-$(1, 1)$-invex function with respect to the same function η and with respect to the function b_g defined by

$$b_g(x, \overline{x}) = x^2 + x + e^x.$$

Further, there exist the Lagrange multipliers $\overline{\lambda} = \left[1, \frac{1}{2(1+e)} \right] \in R_+^2$ and $\overline{\xi} = 1 \in R_+$ such that the Karush-Kuhn-Tucker necessary optimality conditions are satisfied at $\overline{x}$. Since all hypotheses of Theorem 3 are satisfied, then $\overline{x} = 0$ is a properly efficient solution in the considered multiobjective programming problem (VP). Note, moreover, that the functions constituting the considered multiobjective programming problem (VP) are neither convex nor invex at $\overline{x}$ on D (with respect to the function η). Therefore, the results established in [11], [12], [24] and [26] are not applicable in this case.

In the next theorem, we invoke the strict B-(p,r)-invexity restriction at $\overline{x}$ on the Lagrange function f, to obtain that $\overline{x}$ is a properly efficient solution in problem (VP).

Theorem 4 *Let $\overline{x}$ be a feasible point for (VP) and there exist the Lagrange multipliers $\overline{\lambda} \in R_+^k$ and $\overline{\xi} \in R_+^m$ such that the Karush-Kuhn-Tucker necessary optimality conditions are satisfied at $\overline{x}$. Further, assume that the vector-valued Lagrange function is strictly B-(p,r)-invex with respect to η at $\overline{x}$ on D. Then $\overline{x}$ is a properly efficient solution in (VP).*

Proof. Assume that $\overline{x}$ is a feasible point for (VP) and there exist the Lagrange multipliers $\overline{\lambda} \in R_+^k$ and $\overline{\xi} \in R_+^m$ such that the Karush-Kuhn-Tucker necessary optimality conditions are satisfied at $\overline{x}$. We proceed by contradiction. Suppose that $\overline{x}$ is not an efficient solution in problem (VP). Then, there exists a feasible solution $\widetilde{x}$ in (VP) such that

$$f\left(\widetilde{x}\right) \leq f\left(\overline{x}\right). \tag{19}$$

By the Karush-Kuhn-Tucker necessary optimality conditions (8), we get

$$diag\,\overline{\lambda}f\left(\widetilde{x}\right) \leq diag\,\overline{\lambda}f\left(\overline{x}\right). \tag{20}$$

By assumption, the vector-valued Lagrange function L_k is strictly B-(p,r)-invex with respect to η at $\overline{x}$ on D. Then, by Definition 1, it follows that the inequality

$$\frac{1}{r}diag\,b\left(x,\overline{x}\right)\left(e^{r\left(L_k(x,\overline{\lambda},\overline{\xi})-L_k(\overline{x},\overline{\lambda},\overline{\xi})\right)} - 1\right) > \frac{1}{p}\nabla L_k(\overline{x},\overline{\lambda},\overline{\xi})\left(e^{p\eta(x,\overline{x})} - \mathbf{1}\right)$$

holds for all $x \in D$. Therefore, it is also satisfied for $x = \widetilde{x}$. Hence, by definition of the Lagrange function,

$$\frac{1}{r}diag\,b\left(\widetilde{x},\overline{x}\right)\left(e^{r\left(diag\,\overline{\lambda}f(\widetilde{x})+\frac{1}{k}\overline{\xi}g(\widetilde{x})\mathbf{e}-diag\,\overline{\lambda}f(\overline{x})-\frac{1}{k}\overline{\xi}g(\overline{x})\mathbf{e}\right)} - 1\right) >$$
$$\frac{1}{p}\left(diag\,\overline{\lambda}\nabla f(\overline{x}) + \frac{1}{k}\overline{\xi}\nabla g\left(\overline{x}\right)\mathbf{e}\right)\left(e^{p\eta(\widetilde{x},\overline{x})} - \mathbf{1}\right). \tag{21}$$

Therefore, for any $i \in I$,

$$\frac{1}{r}b_i\left(\widetilde{x},\overline{x}\right)\left(e^{r\left(\overline{\lambda}_if_i(\widetilde{x})+\frac{1}{k}\overline{\xi}g(\widetilde{x})-\overline{\lambda}_if_i(\overline{x})-\frac{1}{k}\overline{\xi}g(\overline{x})\right)} - 1\right) >$$
$$\frac{1}{p}\left(\overline{\lambda}_i\nabla f_i(\overline{x}) + \frac{1}{k}\overline{\xi}\nabla g\left(\overline{x}\right)\right)\left(e^{p\eta(\widetilde{x},\overline{x})} - \mathbf{1}\right).$$

Adding both sides of the above inequalities, we obtain

$$\frac{1}{r}\sum_{i=1}^{k}b_i\left(\widetilde{x},\overline{x}\right)\left(e^{r\left(\overline{\lambda}_if_i(\widetilde{x})+\frac{1}{k}\overline{\xi}g(\widetilde{x})-\overline{\lambda}_if_i(\overline{x})-\frac{1}{k}\overline{\xi}g(\overline{x})\right)} - 1\right) >$$
$$\frac{1}{p}\left(\overline{\lambda}\nabla f(\overline{x}) + \overline{\xi}\nabla g\left(\overline{x}\right)\right)\left(e^{p\eta(\widetilde{x},\overline{x})} - \mathbf{1}\right). \tag{22}$$

Using the feasibility of $\widetilde{x}$ in (VP) together with the Karush-Kuhn-Tucker necessary optimality conditions (7) and (8), we get

$$\overline{\xi}g(\widetilde{x}) \leqq \overline{\xi}g(\overline{x}). \tag{23}$$

From the definition of strictly $B - (p, r)$-invexity, it follows that $b(\widetilde{x}, \overline{x}) \geqq 0$. Thus, by (20) and (23),

$$\frac{1}{r} \sum_{i=1}^{k} b_i (\widetilde{x}, \overline{x}) \left(e^{r\left(\overline{\lambda}_i f_i(\widetilde{x}) + \frac{1}{k}\overline{\xi}g(\widetilde{x}) - \overline{\lambda}_i f_i(\overline{x}) - \frac{1}{k}\overline{\xi}g(\overline{x})\right)} - 1 \right) < 0. \tag{24}$$

By (22) and (24), we obtain the inequality

$$\frac{1}{p} \left(\overline{\lambda}\nabla f(\overline{x}) + \overline{\xi}\nabla g(\overline{x}) \right) \left(e^{p\eta(\widetilde{x}, \overline{x})} - 1 \right) < 0,$$

which contradicts the Karush-Kuhn-Tucker necessary optimality condition (6). This means that $\overline{x}$ is an efficient solution in (VP). The last part of the proof is similar as the proof of Theorem 3. ∎

4   Vector Mond-Weir duality

Now, we consider the following multiobjective dual problem:

$$V - \text{maximize} \quad f(y) = (f_1(y), f_2(y), ..., f_k(y))$$

$$\text{(VMWD)} \quad \begin{aligned} \frac{1}{p} \left(\lambda\nabla f(y) + \xi\nabla g(y) \right) \left(e^{p\eta(x,y)} - \mathbf{1} \right) \geqq 0, & \quad \text{if} \quad p \neq 0 \\ \left(\lambda\nabla f(y) + \xi\nabla g(y) \right) \eta(x, y) \geqq 0, & \quad \text{if} \quad p = 0 \end{aligned} \quad \forall x \in D,$$

$$\xi g(y) \geqq 0,$$

$$y \in X, \ \lambda \in R_+^k, \ \lambda > 0, \ \lambda^T \mathbf{e} = 1, \ \xi \in R_+^m, \ \xi \geqq 0,$$

where $\mathbf{e} = (1, ..., 1) \in R^k$.

It is worth to notice that the above dual for the considered multiobjective programming problem is known as Mond-Weir [21] type dual. Duality theorems for this kind of dual pairs in the differentiable case have not been examined in a vectorial case yet.

Let W be the set of all feasible solutions in (VMWD) and $pr_X W$ be the projection of the set W on X.

Efficient points and properly efficient points for the considered vector Mond-Weir dual problem (VMWD) are defined in a manner analogous to those of the primal multiobjective programming problem (VP) by simply reversing the inequalities in Definitions 3 and 4.

Now, we give some useful lemma whose a simple proof will be omitted in the paper.

Lemma 5 *Let* (y, λ, ξ) *be a feasible point in (VMWD). If* g_j, $j \in J(y)$, *are* b_{g_j}-(p, r)-*invex with respect to the same function* η *at* y *on* $D \cup pr_X W$, *then, for all* $x \in D$, *the following inequality is satisfied:*

$$\begin{aligned} \frac{1}{p}\xi\nabla g(y) \left(e^{p\eta(x,y)} - \mathbf{1} \right) \leqq 0, & \quad \text{if} \quad p \neq 0, \\ \xi\nabla g(y)\eta(x, y) \leqq 0, & \quad \text{if} \quad p = 0. \end{aligned} \tag{25}$$

Theorem 6 *(Weak duality): Let x and (y, λ, ξ) be feasible points for (VP) and (VMWD), respectively. Further, assume that f_i, $i \in I$, are b_{f_i}-(p, r)-invex with respect to the same function η at y on $D \cup pr_X W$, $b_{f_i}(x, y) > 0$, $i \in I$, and, moreover, g_j, $j \in J(y)$, are b_{g_j}-(p, r)-invex with respect to the same function η at y on $D \cup pr_X W$. Then $f(x) \nleq f(y)$.*

Proof. Suppose, contrary to what we desire to show, that $f(x) \leq f(y)$. Since f_i, $i \in I$, are b_{f_i}-(p, r)-invex with respect to the same function η at y on $D \cup pr_X W$, it follows, by Proposition 1 b), that any function f_i, $i \in I$, is b_{f_i}-$(p, \frac{r}{\lambda_i})$-invex with respect to the same function η at y on $D \cup pr_X W$. Hence, by Definition 2, the following inequality

$$\frac{\lambda_i}{r} b_{f_i}(x, y) \left(e^{\frac{r}{\lambda_i} \lambda_i [f_i(x) - f_i(y)]} - 1 \right) \geqq \frac{1}{p} \lambda_i \nabla f_i(y) \left(e^{p\eta(x,y)} - 1 \right).$$

holds for any $i \in I$. Hence, using $f(x) \leq f(y)$ together with the assumption imposed on a function b_f, we get, for any $i \in I$,

$$\frac{1}{p} \lambda_i \nabla f_i(y) \left(e^{p\eta(x,y)} - 1 \right) \leqq 0,$$

and for at least one $i_0 \in I$,

$$\frac{1}{p} \lambda_{i_0} \nabla f_{i_0}(y) \left(e^{p\eta(x,y)} - 1 \right) < 0.$$

After adding both sides of the above inequalities, we obtain

$$\frac{1}{p} \lambda \nabla f(y) \left(e^{p\eta(x,y)} - 1 \right) < 0. \tag{26}$$

By assumption, g_j, $j \in J(y)$, are b_{g_j}-(p, r)-invex with respect to the same function η at y on $D \cup pr_X W$. Then, by Lemma 5, the inequality (25) holds. After adding both sides of (25) and (26), we obtain that the following inequality

$$\frac{1}{p} \left(\lambda \nabla f(y) + \xi \nabla g(y) \right) \left(e^{p\eta(x,y)} - 1 \right) < 0,$$

which is a contradiction to the first constraint of the vector Mond-Weir dual problem (VMWD). ∎

Theorem 7 *(Strong duality). Let $\overline{x}$ be a properly efficient solution in (VP) and the generalized Slater constraint qualification (GSQC) be satisfied at $\overline{x}$. Then there exist $\overline{\lambda} > 0$ and $\overline{\xi} \geqq 0$ such that $(\overline{x}, \overline{\lambda}, \overline{\xi})$ is feasible in (VMWD) with $\overline{\xi} g(\overline{x}) = 0$. If also weak duality (Theorem 6) holds, then $(\overline{x}, \overline{\lambda}, \overline{\xi})$ is a properly efficient solution in the Mond-Weir dual problem (VMWD), and the optimal objective function values are equal in problems (VP) and (VMWD).*

Proof. Since $\overline{x}$ satisfies all hypotheses of Theorem 2, there exist the Lagrange multipliers $\overline{\lambda} > 0$ and $\overline{\xi} \geq 0$ such that the Karush-Kuhn-Tucker conditions (6)-(8) are satisfied. Hence, $\left(\overline{x}, \overline{\lambda}, \overline{\xi}\right)$ is feasible in (VMWD). Thus, by weak duality (Theorem 6), it follows that $\left(\overline{x}, \overline{\lambda}, \overline{\xi}\right)$ is an efficient solution in (VMWD).

We shall prove that $\left(\overline{x}, \overline{\lambda}, \overline{\xi}\right)$ is a properly efficient solution in (VMWD) by the method of contradiction. Suppose that $\left(\overline{x}, \overline{\lambda}, \overline{\xi}\right)$ is not so. Then, there exists $\left(\widetilde{y}, \widetilde{\lambda}, \widetilde{\xi}\right)$ feasible in (VMWD) and an objective function f_s satisfying the following inequality

$$f_s\left(\widetilde{y}\right) - f_s\left(\overline{x}\right) > M\left[f_j\left(\widetilde{y}\right) - f_j\left(\overline{x}\right)\right] \tag{27}$$

for every scalar $M > 0$ and all j satisfying

$$f_j\left(\widetilde{y}\right) > f_j\left(\overline{x}\right). \tag{28}$$

We divide the index set I and denote by I_1 the set of indexes of objective functions satisfying the inequality (28). By I_2 we denote the set of indexes of objective functions defining by $I_2 = I \setminus (I_1 \cup s)$. Let $M > \frac{\lambda_j}{\lambda_i} |I_1|$, where $|I_1|$ denotes the number of elements in the set I_1. Hence, by (27) and (28),

$$\lambda_s\left(f_s\left(\widetilde{y}\right) - f_s\left(\overline{x}\right)\right) > \sum_{j \in I_1} \lambda_j\left[f_j\left(\widetilde{y}\right) - f_j\left(\overline{x}\right)\right]. \tag{29}$$

Using the definition of the set I_2 together with (29), we obtain

$$\sum_{i=1}^{k} \lambda_i f_i\left(\overline{x}\right) \quad - \quad \lambda_s f_s\left(\overline{x}\right) + \sum_{j \in I_1} \lambda_j f_j\left(\overline{x}\right) + \sum_{j \in I_2} \lambda_j f_j\left(\overline{x}\right)$$

$$< \quad \lambda_s f_s\left(\widetilde{y}\right) + \sum_{j \in I_1} \lambda_j f_j\left(\widetilde{y}\right) + \sum_{j \in I_2} \lambda_j f_j\left(\widetilde{y}\right) = \sum_{i=1}^{k} \lambda_i f_i\left(\widetilde{y}\right).$$

This is a contradiction to the weak duality theorem. Hence, $\left(\overline{x}, \overline{\lambda}, \overline{\xi}\right)$ is a properly efficient solution in the vector Mond-Weir dual problem (VMWD), and the optimal objective function values in the primal and the dual problems are equal. ∎

Theorem 8 *(Converse duality): Let $\left(\overline{y}, \overline{\lambda}, \overline{\xi}\right)$ be feasible in (VMWD). Further, assume that f_i, $i \in I$, is b_{f_i}-(p, r)-invex with respect to η at $\overline{y}$ on $D \cup pr_X W$ and g_j, $j \in J\left(\overline{y}\right)$, is b_{g_j}-(p, r)-invex at $\overline{y}$ on $D \cup pr_X W$ with respect to the same function η. Then $\overline{y}$ is a properly efficient solution in problem (VP).*

Proof. First, we prove that $\overline{y}$ is an efficient point in (VP). Suppose that $\overline{y}$ is not so. Then, there exists $\widetilde{x}$ feasible in (VP) such that

$$f(\widetilde{x}) \leq f(\overline{y}). \tag{30}$$

By assumption, f_i, $i \in I$, is b_{f_i}-(p, r)-invex with respect to the same function η at $\overline{y}$ on $D \cup pr_X W$ and g_j, $j \in J(\overline{y})$, are b_{g_j}-(p, r)-invex with respect to the same function η at $\overline{y}$ on $D \cup pr_X W$. Since $(\overline{y}, \overline{\lambda}, \overline{\xi})$ is feasible in (VMWD), then, by Proposition 1 b), it follows that any function $\overline{\lambda}_i f_i$, $i \in I$, is b_{f_i}-$(p, \frac{r}{\lambda_i})$-invex with respect to the same function η at $\overline{y}$ on $D \cup pr_X W$, and any function $\overline{\xi}_j g_j$, $j \in J(\overline{y})$, is b_{g_j}-$(p, \frac{r}{\xi_j})$-invex with respect to η at $\overline{y}$ on $D \cup pr_X W$. Then, by Definition 2, the following inequalities

$$\frac{\overline{\lambda}_i}{r} b_{f_i}(x, \overline{y}) \left(e^{\frac{r}{\lambda_i} \overline{\lambda}_i [f_i(x) - f_i(\overline{y})]} - 1 \right) \geqq \frac{1}{p} \overline{\lambda}_i \nabla f_i(\overline{y}) \left(e^{p\eta(x, \overline{y})} - \mathbf{1} \right), \forall i \in I, \tag{31}$$

$$\frac{\overline{\xi}_j}{r} b_{g_j}(x, \overline{y}) \left(e^{\frac{r}{\xi_j} \overline{\xi}_j [g_j(x) - g_j(\overline{y})]} - 1 \right) \geqq \frac{1}{p} \overline{\xi}_j \nabla g_j(\overline{y}) \left(e^{p\eta(x, \overline{y})} - \mathbf{1} \right), \forall j \in J(\overline{y}) \tag{32}$$

hold for all $x \in D$. Thus, by (30) and (31), for $x = \widetilde{x}$,

$$\frac{1}{p} \overline{\lambda}_i \nabla f_i(\overline{y}) \left(e^{p\eta(\widetilde{x}, \overline{y})} - \mathbf{1} \right) \leqq 0, \forall i \in I, \tag{33}$$

$$\frac{1}{p} \overline{\lambda}_s \nabla f_s(\overline{y}) \left(e^{p\eta(\widetilde{x}, \overline{y})} - \mathbf{1} \right) < 0 \text{ for some } s \in I. \tag{34}$$

Then, by (33) and (34), respectively, we get

$$\frac{1}{p} \overline{\lambda} \nabla f(\overline{y}) \left(e^{p\eta(\widetilde{x}, \overline{y})} - \mathbf{1} \right) < 0. \tag{35}$$

Using the feasibility of $\widetilde{x}$ in (VP) and the feasibility of $(\overline{y}, \overline{\lambda}, \overline{\xi})$ in (VMWD), (32) yields

$$\frac{1}{p} \overline{\xi}_j \nabla g_j(\overline{y}) \left(e^{p\eta(\widetilde{x}, \overline{y})} - \mathbf{1} \right) \leqq 0.$$

Taking into account the constraint functions g_j, $j \notin J(\overline{y})$, we get the following inequality

$$\frac{1}{p} \overline{\xi} \nabla g(\overline{y}) \left(e^{p\eta(\widetilde{x}, \overline{y})} - \mathbf{1} \right) \leqq 0. \tag{36}$$

By (35) and (36), we obtain

$$\frac{1}{p} \left[\overline{\lambda} \nabla f(\overline{y}) + \overline{\xi} \nabla g(\overline{y}) \right] \left(e^{p\eta(\widetilde{x}, \overline{y})} - \mathbf{1} \right) < 0,$$

which is a contradiction to the feasibility of $(\overline{y}, \overline{\lambda}, \overline{\xi})$ in (VMWD). This means that $\overline{y}$ is an efficient solution in (VP).

We now show that $\overline{y}$ is a properly efficient solution in (VP). We proceed by contradiction. Suppose that $\overline{y}$ is not a properly efficient point in (VP). Then, by Definition 4, there exist $\widehat{x} \in D$ and an index i such that, for every positive scalar M, the following inequality

$$\frac{f_i(\overline{y}) - f_i(\widehat{x})}{f_j(\widehat{x}) - f_j(\overline{y})} > M, \tag{37}$$

holds for all j satisfying $f_j(\widehat{x}) > f_j(\overline{y})$, whenever $f_i(\widehat{x}) \leqq f_i(\overline{y})$. Since $M(f_j(\widehat{x}) - f_j(\overline{y})) > 0$, whereas $f_i(\widehat{x}) - f_i(\overline{y}) \leqq 0$, then we obtain a contradiction to (37). Hence, $\overline{y}$ is a properly efficient solution in (VP). ∎

5 Wolfe dualtiy with vector constraints

Relative to problem (VP), we consider the following dual problem in the Wolfe format [27]:

$$V-\text{maximize } f(y) + \xi g(y)\mathbf{e}$$

(VWD1)
$$\begin{aligned} \frac{1}{p}\left(diag\,\lambda\nabla f(y) + \frac{1}{k}\xi\nabla g(y)\mathbf{e}\right)\left(e^{p\eta(x,y)} - \mathbf{1}\right) &\geqq 0, \quad \text{if} \quad p \neq 0 \\ \left(diag\,\lambda\nabla f(y) + \frac{1}{k}\xi\nabla g(y)\mathbf{e}\right)\eta(x,y) &\geqq 0, \qquad \text{if} \quad p = 0 \end{aligned} \quad \forall x \in D,$$

$$y \in X, \ \lambda \in R_+^k, \ \lambda > 0, \ \lambda\mathbf{e} = 1, \ \xi \in R_+^m, \ \xi \geqq 0.$$

Let $\widetilde{W_1}$ denote the set of all feasible solutions in the Wolfe dual problem (VWD1).

Theorem 9 *(Weak duality): Let x and (y, λ, ξ) be feasible points for (VP) and (VWD1), respectively. Further, assume that the vector-valued Lagrange function is B-(p, r)-invex with respect to η at y on $D \cup pr_X\widetilde{W_1}$ and $b(x, y) > 0$. Then $f(x) \nleq f(y) + \xi g(y)\mathbf{e}$.*

Proof. We assume that $f(x) \leq f(y) + \xi g(y)\mathbf{e}$ and exhibit a contradiction. Since $x \in D$ and $\xi \geqq 0$, then $\xi g(x) \leqq 0$. Thus,

$$f(x) + \xi g(x)\mathbf{e} \leq f(y) + \xi g(y)\mathbf{e}. \tag{38}$$

Therefore, for any $i \in I$,

$$f_i(x) + \xi g(x) \leqq f_i(y) + \xi g(y), \tag{39}$$

but for at least one $i_0 \in I$,

$$f_{i_0}(x) + \xi g(x) < f_{i_0}(y) + \xi g(y), \tag{40}$$

Multiplying both sides of (39) by λ_i and (40) by λ_{i_0}, we get

$$\lambda_i f_i(x) + \lambda_i \xi g(x) \leqq \lambda_i f_i(y) + \lambda_i \xi g(y), \tag{41}$$

$$\lambda_{i_0} f_{i_0}(x) + \lambda_{i_0}\xi g(x) < \lambda_{i_0}f_{i_0}(y) + \lambda_{i_0}\xi g(y). \tag{42}$$

Adding both sides of (41) and (42), we obtain

$$\sum_{i=1}^{k} \lambda_i f_i(x) + \xi g(x) \sum_{i=1}^{k} \lambda_i < \sum_{i=1}^{k} \lambda_i f_i(y) + \xi g(y) \sum_{i=1}^{k} \lambda_i.$$

From the feasibility of (y, λ, ξ) in (VWD1), it follows that $\lambda\mathbf{e} = 1$. Thus,

$$\sum_{i=1}^{k} \lambda_i f_i(x) + \xi g(x) < \sum_{i=1}^{k} \lambda_i f_i(y) + \xi g(y). \tag{43}$$

By assumption, the vector-valued Lagrange function L_k is B-(p, r)-invex with respect to η on $D \cup pr_X \widetilde{W}_1$. Then, by Definition 1, the inequality

$$\frac{1}{r} diag\, b\,(x, y) \left(e^{r(L_k(x,\lambda,\xi) - L_k(y,\lambda,\xi))} - 1 \right) \geqq \frac{1}{p} \nabla L_k(y, \lambda, \xi) \left(e^{p\eta(x,y)} - 1 \right)$$

holds for all $x \in D \cup pr_X \widetilde{W}_1$. Hence, by the definition of the modified vector-valued Lagrange function L_k,

$$\begin{aligned}\frac{1}{r} diag\, b\,(x, y) \left(e^{r\left(diag\,\lambda f(x) + \frac{1}{k}\xi g(x)\mathbf{e} - diag\,\lambda f(y) - \frac{1}{k}\xi g(y)\mathbf{e}\right)} - 1 \right) \geqq \\ \frac{1}{p} \left(diag\, \lambda \nabla f(y) + \frac{1}{k}\xi \nabla g(y)\, \mathbf{e} \right) \left(e^{p\eta(x,y)} - 1 \right).\end{aligned} \tag{44}$$

From the constraints of (VWD2), it follows that

$$\frac{1}{r} diag\, b\,(x, y) \left(e^{r\left(diag\,\lambda f(x) + \frac{1}{k}\xi g(x)\mathbf{e} - diag\,\lambda f(y) - \frac{1}{k}\xi g(y)\mathbf{e}\right)} - 1 \right) \geqq 0,$$

and, so, for any $i \in I$,

$$\frac{1}{r} b_i\,(x, y) \left(e^{r\left(\lambda_i f_i(x) + \frac{1}{k}\xi g(x) - \lambda_i f_i(y) - \frac{1}{k}\xi g(y)\right)} - 1 \right) \geqq 0, \tag{45}$$

By assumption, $b_i\,(x, y) > 0$ for any $i \in I$. Thus, (45) yields

$$\lambda_i f_i(x) + \frac{1}{k}\xi g(x) - \lambda_i f_i(y) - \frac{1}{k}\xi g(y) \geqq 0.$$

Adding both sides of the inequalities above, we get

$$\sum_{i=1}^{k} \lambda_i f_i(x) + \sum_{i=1}^{k} \frac{1}{k}\xi g(x) \geqq \sum_{i=1}^{k} \lambda_i f_i(y) + \sum_{i=1}^{k} \frac{1}{k}\xi g(y).$$

Hence,

$$\sum_{i=1}^{k} \lambda_i f_i(x) + \xi g(x) \geqq \sum_{i=1}^{k} \lambda_i f_i(y) + \xi g(y),$$

contradicting the inequality (43). ∎

Theorem 10 *(Strong duality). Let $\overline{x}$ be a properly efficient solution in (VP) and the generalized Slater constraint qualification (GSQC) be satisfied at $\overline{x}$. Assume that there exist $\overline{\lambda} \in R_+^k$, $\overline{\xi} \in R_+^m$, such that $(\overline{x}, \overline{\lambda}, \overline{\xi})$ is feasible for (VWD). Then the objective functions of (VP) and (VWD2) are equal at these points. If also weak duality (Theorem 9) between (VP) and (VWD2) holds then $(\overline{x}, \overline{\lambda}, \overline{\xi})$ is a properly efficient solution in (VWD2).*

Proof.   Proof of this theorem is similar to the proof of Theorem 7. ■

Theorem 11 *(Converse duality): Let* $\left(\overline{y},\overline{\lambda},\overline{\xi}\right)$ *be feasible in (VWD1) such that* $\overline{\xi}g\left(\overline{y}\right) = 0$. *Moreover, assume that the Lagrange function is B-(p, r)-invex with respect η at $\overline{y}$ on* $D \cup pr_X \widetilde{W}_2$. *Then $\overline{y}$ is a properly efficient solution in (VP).*

Proof.   Suppose that $\overline{y}$ is not an efficient solution in (VP). Then, there exists $\widetilde{x} \in D$ such that

$$f\left(\widetilde{x}\right) \leq f\left(\overline{y}\right). \tag{46}$$

By assumption, $\overline{\xi}g\left(\overline{y}\right) = 0$. Using the feasibility $\widetilde{x}$ in (VP) together with $\overline{\xi} \geq 0$, we obtain

$$\overline{\xi}g\left(\widetilde{x}\right) \leqq 0. \tag{47}$$

Since $\overline{\lambda} \in R_+^k$ and $\overline{\lambda}\mathbf{e} = 1$, then by (46) and (47),

$$diag\,\overline{\lambda}f\left(\widetilde{x}\right) + \frac{1}{k}\overline{\xi}g\left(\widetilde{x}\right)\mathbf{e} \leq diag\,\overline{\lambda}f\left(\overline{y}\right) + \frac{1}{k}\overline{\xi}g\left(\overline{y}\right)\mathbf{e}. \tag{48}$$

By assumption, the modified vector-valued Lagrange function is B-(p, r)-invex with respect η at $\overline{y}$ for all $x \in D \cup pr_X \widetilde{W}_2$. Hence, by definition of the modified vector-valued Lagrange function,

$$\frac{1}{r}b\left(\widetilde{x},\overline{x}\right)\left(e^{r\left(diag\,\overline{\lambda}f(\widetilde{x})+\frac{1}{k}\overline{\xi}g(\widetilde{x})\mathbf{e}-diag\,\overline{\lambda}f(\overline{y})-\frac{1}{k}\overline{\xi}g(\overline{y})\mathbf{e}\right)} - 1\right) \geqq \\ \frac{1}{p}\left(diag\,\overline{\lambda}\nabla f\left(\overline{y}\right) + \frac{1}{k}\overline{\xi}\nabla g\left(\overline{y}\right)\mathbf{e}\right)\left(e^{p\eta(\widetilde{x},\overline{y})} - 1\right). \tag{49}$$

Using the constraint of (VWD1), we get

$$\frac{1}{r}b\left(\widetilde{x},\overline{x}\right)\left(e^{r\left(diag\,\overline{\lambda}f(\widetilde{x})+\frac{1}{k}\overline{\xi}g(\widetilde{x})\mathbf{e}-diag\,\overline{\lambda}f(\overline{y})-\frac{1}{k}\overline{\xi}g(\overline{y})\mathbf{e}\right)} - 1\right) \geqq 0.$$

Since $b\left(\widetilde{x},\overline{x}\right) > 0$, then, by (49), it follows that

$$diag\,\overline{\lambda}f(\widetilde{x}) + \frac{1}{k}\overline{\xi}g(\widetilde{x})\mathbf{e} \geqq diag\,\overline{\lambda}f(\overline{y}) + \frac{1}{k}\overline{\xi}g(\overline{y})\mathbf{e}$$

contradicting the inequality (48). The proof of properly efficiency of $\overline{y}$ in (VP) is similar as in the proof of Theorem 8. ■

6   Wolfe duality with scalar constraints

We now consider the following dual problem in the Wolfe format [27]:

$$V-\text{maximize } f(y) + \xi g(y)\mathbf{e}$$

$$\text{(VWD2)} \quad \begin{array}{l} \frac{1}{p}\left(\lambda\nabla f(y) + \xi\nabla g(y)\right)\left(e^{p\eta(x,y)} - \mathbf{1}\right) \geqq 0, \quad \text{if} \quad p \neq 0 \\ \left(\lambda\nabla f(y) + \xi\nabla g(y)\right)\eta(x,y) \geqq 0, \qquad \text{if} \quad p = 0 \end{array} \quad \forall x \in D,$$

$$y \in X, \ \lambda \in R_+^k, \ \lambda > 0, \ \lambda\mathbf{e} = 1, \ \xi \in R_+^m, \ \xi \geqq 0.$$

Let $\widetilde{W}_2$ denote the set of all feasible points of (VWD2).

Theorem 12 *(Weak duality): Let x and (y, λ, ξ) be feasible solutions for (VP) and (VWD2), respectively. Assume that one of the following hypotheses:*

   i) the vector-valued Lagrange function L_k is B-(p,r)-invex with respect to η at y on $D \cup pr_X \widetilde{W}_2$ and $b(x,y) > 0$,

   ii) the scalar Lagrange function L is B-(p,r)-invex with respect to η at y on $D \cup pr_X \widetilde{W}_2$ and $b(x,y) > 0$,

   iii) λf is B-(p,r)-invex at y on $D \cup pr_X \widetilde{W}_2$ with respect to η, $-\xi g$ is B-(p,r)-incave at y on $D \cup pr_X \widetilde{W}_2$ with respect to the same function η and $b(x,y) > 0$,

   is satisfied. Then $f(x) \not\leq f(y) + \xi g(y)\mathbf{e}$.

Proof. Assume that

$$f(x) \leq f(y) + \xi g(y)\mathbf{e} \tag{50}$$

and exhibit a contradiction. Since $x \in D$ and $\xi \geqq 0$, then $\xi g(x) \leqq 0$. Hence, (50) yields, for any $i \in I$,

$$f_i(x) + \xi g(x) \leqq f_i(y) + \xi g(y).$$

and for at least one $i_0 \in I$,

$$f_{i_0}(x) + \xi g(x) \leqq f_{i_0}(y) + \xi g(y).$$

Multiplying the above inequalities by $\lambda_i > 0$, we get

$$\lambda_i f_i(x) + \lambda_i \xi g(x) \leqq \lambda_i f_i(y) + \lambda_i \xi g(y), \quad i \in I, \tag{51}$$

and for at least one $i_0 \in I$,

$$\lambda_{i_0} f_{i_0}(x) + \lambda_{i_0} \xi g(x) < \lambda_{i_0} f_{i_0}(y) + \lambda_{i_0} \xi g(y). \tag{52}$$

i) Assume that the vector-valued Lagrange function L_k is B-(p,r)-invex with respect to η at y on $D \cup pr_X \widetilde{W}_2$. Hence, by Definition 1,

$$\frac{1}{r} b(x,y) \left(e^{r\left(diag\, \lambda f(x) + \frac{1}{k}\xi g(x)\mathbf{e} - diag\, \lambda f(y) - \frac{1}{k}\xi g(y)\mathbf{e} \right)} - 1 \right) \geqq$$
$$\frac{1}{p} \left(diag\, \lambda \nabla f(y) + \frac{1}{k}\xi \nabla g(y)\, \mathbf{e} \right) \left(e^{p\eta(x,y)} - 1 \right).$$

Since $b(x,y) > 0$, then, using (51) and (52), we obtain

$$\frac{1}{p} \left(diag\, \lambda \nabla f(y) + \frac{1}{k}\xi \nabla g(y)\, \mathbf{e} \right) \left(e^{p\eta(x,y)} - 1 \right) \leq 0.$$

Adding both sides of above inequalities, we obtain the inequality

$$\frac{1}{p}\left(\lambda\nabla f(y) + \xi\nabla g(y)\right)\left(e^{p\eta(x,y)} - \mathbf{1}\right) < 0, \tag{53}$$

which contradicts the dual constraint of (VWD2).

ii) Assume now that the scalar Lagrange function L is B-(p,r)-invex with respect to η at y on $D \cup pr_X\widetilde{W}_2$. Then, by Definition 1,

$$\frac{1}{r}b(x,y)\left(e^{r(\lambda f(x)+\xi g(x)-\lambda f(y)-\xi g(y))} - 1\right) \geqq \frac{1}{p}\left(\lambda\nabla f(y) + \xi\nabla g(y)\right)\left(e^{p\eta(x,y)} - \mathbf{1}\right) \tag{54}$$

Adding both sides of (51) and (52), we get

$$\sum_{i=1}^{k}\lambda_i f_i(x) + \xi g(x)\sum_{i=1}^{k}\lambda_i < \sum_{i=1}^{k}\lambda_i f_i(y) + \xi g(y)\sum_{i=1}^{k}\lambda_i. \tag{55}$$

From the constraints of (VWD2), it follows that $\sum_{i=1}^{k}\lambda_i = 1$. Therefore, (55) gives

$$\lambda f(x) + \xi g(x) < \lambda f(y) + \xi g(y). \tag{56}$$

Since $b(x,y) > 0$, then, using (54) and (56), we obtain (53), contradicting the dual constraint of (VWD2).

iii) Assume that the functions λf is B-(p,r)-invex at y on $D \cup pr_X\widetilde{W}_2$ with respect to η and $-\xi g$ is B-(p,r)-incave at y on $D \cup pr_X\widetilde{W}_2$ with respect to the same function η. By Definition 1, it follows that

$$\frac{1}{r}b(x,y)\left(e^{r(\lambda f(x)-\lambda f(y))} - 1\right) \geqq \frac{1}{p}\lambda\nabla f(y)\left(e^{p\eta(x,y)} - \mathbf{1}\right), \tag{57}$$

$$\frac{1}{r}b(x,y)\left(e^{r(-\xi g(x)-(-\xi g(y)))} - 1\right) \leqq -\frac{1}{p}\xi\nabla g(y)\left(e^{p\eta(x,y)} - \mathbf{1}\right). \tag{58}$$

In similar manner as in the proof under hypothesis ii), we obtain the inequality (56). Hence, by (56),

$$\lambda f(x) - \lambda f(y) < -\xi g(x) - (-\xi g(y)).$$

Since $b(x,y) > 0$, then

$$\frac{1}{r}b(x,y)\left(e^{r(\lambda f(x)-\lambda f(y))} - 1\right) < \frac{1}{r}b(x,y)\left(e^{r(-\xi g(x)-(-\xi g(y)))} - 1\right). \tag{59}$$

Using (57) and (58) together with (59), we obtain

$$\frac{1}{p}\lambda\nabla f(y)\left(e^{p\eta(x,y)} - \mathbf{1}\right) < -\frac{1}{p}\xi\nabla g(y)\left(e^{p\eta(x,y)} - \mathbf{1}\right),$$

and, so

$$\frac{1}{p}\left(\lambda\nabla f(y) + \xi\nabla g(y)\right)\left(e^{p\eta(x,y)} - \mathbf{1}\right) < 0,$$

contradicting the dual constraint of (VWD2). ∎

Theorem 13 *(Strong duality). Let $\overline{x}$ be a properly efficient solution in (VP) and the generalized Slater constraint qualification (GSQC) be satisfied at $\overline{x}$. Then there exist $\overline{\lambda} \in R_+^k, \overline{\xi} \in R_+^m$, such that $\left(\overline{x}, \overline{\lambda}, \overline{\xi}\right)$ is feasible for (VWD2) and the objective functions of (VP) and (VWD2) are equal at these points. If also weak duality (Theorem 12) between (VP) and (VWD2) holds, then $\left(\overline{x}, \overline{\lambda}, \overline{\xi}\right)$ is a properly efficient solution in (VWD2).*

Proof.   Proof of this theorem is similar to the proof of Theorem 7. ■

Theorem 14 *(Converse duality): Let $\left(\overline{y}, \overline{\lambda}, \overline{\xi}\right)$ be feasible in (VWD) such that $\overline{\xi} g\left(\overline{y}\right) = 0$. Moreover, we assume that the vector-valued Lagrange function L_k is B-(p, r)-invex with respect η at $\overline{y}$ on $D \cup pr_X \widetilde{W}_2$. Then $\overline{y}$ is a properly efficient solution in (VP).*

Proof.   Suppose that $\overline{y}$ is not an efficient solution in (VP). Then, there exists $\widetilde{x} \in D$ such that

$$f\left(\widetilde{x}\right) \leq f\left(\overline{y}\right). \tag{60}$$

From the feasibility $\widetilde{x}$ in (VP) together with $\overline{\xi} \geq 0$, it follows that

$$\overline{\xi} g\left(\widetilde{x}\right) \leqq 0. \tag{61}$$

Since $\overline{\lambda} \in R_+^k$ and $\overline{\lambda} e = 1$, then, using (60) and (61) together with assumption $\overline{\xi} g\left(\overline{y}\right) = 0$, we obtain

$$diag\, \overline{\lambda} f\left(\widetilde{x}\right) + \overline{\xi} g\left(\widetilde{x}\right) \mathbf{e} \leq diag\, \overline{\lambda} f\left(\overline{y}\right) + \overline{\xi} g\left(\overline{y}\right) \mathbf{e}. \tag{62}$$

By assumption, the modified vector-valued Lagrange function is B-(p, r)-invex with respect η at $\overline{y}$ for all $x \in D \cup pr_X \widetilde{W}_2$. Hence, by the definition of the modified vector-valued Lagrange function,

$$b\left(\widetilde{x}, \overline{x}\right) \left(e^{r\left(diag\, \overline{\lambda} f(\widetilde{x}) + \frac{1}{k} \overline{\xi} g(\widetilde{x}) \mathbf{e} - diag\, \overline{\lambda} f(\overline{y}) - \frac{1}{k} \overline{\xi} g(\overline{y}) \mathbf{e}\right)} - 1\right) >$$
$$\frac{1}{p} \left(diag\, \overline{\lambda} \nabla f(\overline{y}) + \frac{1}{k} \overline{\xi} \nabla g\left(\overline{y}\right) \mathbf{e}\right) \left(e^{p\eta(\widetilde{x}, \overline{y})} - \mathbf{1}\right).$$

Since $b\left(\widetilde{x}, \overline{x}\right) > 0$, then

$$\frac{1}{p} \left(diag\, \overline{\lambda} \nabla f(\overline{y}) + \frac{1}{k} \overline{\xi} \nabla g(\overline{y})\right) \left(e^{p\eta(\widetilde{x}, \overline{y})} - \mathbf{1}\right) < 0,$$

and, for any $i \in I$,

$$\frac{1}{p} \left(\overline{\lambda}_i \nabla f_i(\overline{y}) + \frac{1}{k} \overline{\xi} \nabla g(\overline{y})\right) \left(e^{p\eta(\widetilde{x}, \overline{y})} - \mathbf{1}\right) < 0.$$

After adding both sides of the above inequalities, we get the following inequality

$$\frac{1}{p} \left(\overline{\lambda} \nabla f(\overline{y}) + \overline{\xi} \nabla g(\overline{y})\right) \left(e^{p\eta(\widetilde{x}, \overline{y})} - \mathbf{1}\right) < 0,$$

which contradicts the dual constraints of problem (VWD). The proof of properly efficiency of $\overline{y}$ in (VP) is similar as in the proof of Theorem 8. ■

References

[1] T. Antczak, (p, r)-invex sets and functions, Journal of Mathematical Analysis and Applications 263 (2001) 355–379.

[2] T. Antczak, A class of B-(p, r)-invex functions and mathematical programming, Journal of Mathematical Analysis and Applications 286 (2003) 187–206.

[3] T. Antczak, (p, r)-invexity in multiobjective programming, European Journal of Operational Research 152 (2004) 72–87.

[4] T. Antczak, r-pre-invexity and r-invexity in mathematical programming, Computers and Mathematics with Applications 50 (2005), 551–566.

[5] M. Avriel, r-convex functions, Mathematical Programming 2 (1972) 309–323.

[6] C.R. Bector, S.K. Suneja and C.S. Lalitha, Generalized B-vex functions and generalized B-vex programming, Journal of Optimization Theory and Applications 76 (1993) 561–576.

[7] A. Ben-Israel and B. Mond, What is invexity?, Journal of Australian Mathematical Society Ser. B 28 (1986) 1–9.

[8] B.D. Craven, Invex functions and constrained local minima, Bulletin of the Australian Mathematical Society 24 (1981) 357–366.

[9] B.D. Craven and B.M. Glover, Invex functions and duality, Journal of Australian Mathematical Society Ser. A 39 (1985) 1 20.

[10] L.N. Das and S. Nanda, Proper efficiency conditions and duality for multiobjective programming problems involving semilocally invex functions, Optimization 34 (1995), 43–51.

[11] R.R. Egudo and M.A. Hanson, Multiobjective duality with invexity, Journal of Mathematical Analysis and Applications 126 (1987) 469–477.

[12] M.A. Geoffrion, Proper efficiency and the theory of vector maximization, Journal of Mathematical Analysis and Applications 22 (1968) 618–630.

[13] M.A. Hanson, On sufficiency of the Kuhn-Tucker conditions, Journal of Mathematical Analysis and Applications 80 (1981) 545–550.

[14] M.A. Hanson and B. Mond, Necessary and sufficient conditions in constrained optimization, Mathematical Programming 37 (1987) 51–58.

[15] R. Hartley, Vector and parametric programming, Journal of Operations Research Society 36 (1985) 423–432.

[16] M.I. Henig, Proper efficiency with respect to the cones, Journal of Optimization Theory and Applications 36 (1982) 387–407.

[17] V. Jeyakumar and B. Mond, On generalized convex mathematical programming, Journal of Australian Mathematical Society Ser. B 34 (1992) 43–53.

[18] R.N. Kaul, S.K. Suneja and S. Srivastava, Optimality criteria and duality in multiple objective optimization involving generalized invexity, Journal of Optimization Theory and Applications 80 (1994) 465–482.

[19] H.W. Kuhn and A.W. Tucker, Nonlinear programming, Proceedings Second Berkeley Symposium on Mathematical Statistics and Probability, University of California Press, Berkeley, California, 1951.

[20] D.H. Martin, The essence of invexity, Journal of Optimization Theory and Applications 47 (1985) 65–76.

[21] B. Mond and T. Weir, Generalized concavity and duality, in:S. Schaible and W.T. Ziemba, Eds., Generalized Concavity in Optimization and economics, pp.263–279, Academic Press, New York, 1981.

[22] V. Pareto, Cours de Economie Politique, Rouge, Lausanne, Switzerland, 1896.

[23] T. Taninio and Y. Sawaragi, Duality theory in multiobjective programming, Journal of Optimization Theory and Applications 27 (1980) 509–529.

[24] T. Weir, Proper efficiency and duality for vector valued optimization problems, Journal of Australian Mathematical Society Ser. A 43 (1987) 24–34.

[25] T. Weir and B. Mond, Pre-invex functions in multiple objective optimization, Journal of Mathematical Analysis and Applications 136 (1988) 29–38.

[26] D.G. White, Concepts of proper efficiency, European Journal of Operational Research 13 (1983) 180–188.

[27] P. Wolfe, A duality theorem for nonlinear programming, Quarterly of Applied Mathematics 19 (1961) 239–244.

CHAPTER 6

On nonsmooth constrained optimization involving generalized type-I conditions.

S. K. Mishra[*]   J. S. Rautela[†]   Sanjay Oli[‡]

Abstract

The goal of this paper is to obtain Kuhn-Tucker type necessary and sufficient optimality conditions for nonsmooth constrained optimization problems involving generalized type-I functions on the objective and constraint functions involved in the problem. Further we study the duality results in the presence of the aforesaid weaker assumptions.

Keywords: Type-I functions; Duality; Nonsmooth Constrained optimization.

1 Introduction

In this paper, we consider the following constrained optimization problem

$$\textbf{(P)} \quad \inf_{x \in A} \ f(x), \text{ where } A = \{x \in X : g_j(x) \leq 0, \ j = 1, 2, ..., m\},$$

where f and g_j are nondifferentiable functions.

Hanson [4] introduced the concept of invexity in constrained optimization as a generalization of convexity. Later Craven and Glover [3] proved that any differentiable scalar function is invex if and only if every stationary point is a global minimizer. Therefore in the case of unconstrained optimization, the concept of invexity is equivalent to the above property. However, in constrained optimization the objective and the constraint functions are assumed to be invex with respect to the same vector function. Several authors have considered possible relaxation of the invexity requirement in order to obtain necessary and sufficient conditions for optimality and duality theorems.

Hanson and Mond [5] introduced two new classes of functions called type-I and type-II functions, which are not only sufficient but are also necessary for optimality in the primal and dual problems, respectively. Rueda and Hanson [14] further extended type-I

[*]Department of Mathematics, Faculty of Science, Banaras Hindu University, Varanasi- 221005, India. e-mail: shashikmishra@rediffmail.com

[†]Department of Mathematics, Faculty of Applied Science and Humanities, Echelon Institute of Technology, Faridabad 121001, India. e-mail: sky_dreamz@rediffmail.com

[‡]Department of Mathematics, Asia Pacific Institute of Information Technology SD India. e-mail: sanjayoli@rediffmail.com

functions to pseudo-type-I and quasi-type-I functions and obtained sufficient optimality results for a nonlinear optimization problem involving these functions. Kaul, Suneja and Srivastava [6] obtained sufficient optimality and duality results for a multipleobjective optimization problem under various weak invexity assumptions, such as pseudo-quasi-type-I and quasi-pseudo-type-I functions. For more details one can consult [9-13]. Clarke [2] introduced the concept of subdifferential functions. Recently Caristi, Ferrara and Stefanescu [1] introduced the concepts of η-inf-invex, η-sup-invex, η-inf-pseudo-invex and η-sup-quasi-invex as an extension of invexity.

Motivated by the work of Caristi, Ferrara and Stefanescu [1], we introduce six new classes of type-I functions and establish Kuhn-Tucker type necessary and sufficient optimality conditions and various duality theorems. The results of Hanson [4], Hanson and Mond [5], Rueda and Hanson [14], Kaul, Suneja and Srivastava [6] and Caristi, Ferrara and Stefanescu [1] are special cases of the present work.

2 Preliminaries

Firstly, we recall some known results and concepts. Let X be a nonempty subset of $\mathbb{R}^n$.

Definition 1 *A function $f : X \to \mathbb{R}$ is said to be Lipschitz near $x \in X$ if for some $K > 0$, $|f(y) - f(z)| \leq K \|y - z\|$ for all y, z within a neighborhood of x.*

We say that $f : X \to \mathbb{R}$ is locally Lipschitz on X if it is Lipschitz near any point of X.

If $f : X \to \mathbb{R}$ is Lipschitz at $x \in X$, the generalized derivative (in the sense of Clarke) of f at $x \in X$ in the direction $\mu \in \mathbb{R}^n$, denoted by $f^\circ(x; \mu)$, is given by

$$f^\circ(x_0, \mu) = \lim_{y \to x} \sup_{\lambda \downarrow 0} \frac{f(y + \lambda\mu) - f(y)}{\lambda}.$$

The Clarkes generalized gradient of f at $x \in X$, denoted by $\partial f(x)$, is defined as follows:

$$\partial f(x) = \left\{ x \in R^n : f^\circ(x; \mu) > \xi^T \mu, \ \forall \ \mu \in \mathbb{R}^n \right\}.$$

It follows that, $f^\circ(x; \mu) = \max \left\{ \xi^T \mu : \xi \in \partial f(x) \right\}$, for any $\mu \in \mathbb{R}^n$.

Lemma 1 *For a nondifferentiable constrained optimization problem, a Kuhn-Tucker point is a pair $(x_0, \nu) \in A \times R^m_+$ satisfying the following two conditions:*

$$\langle \xi, \eta(x, x_0) \rangle + \sum_{j=1}^m \nu_j \ \langle \zeta_j, \eta(x, x_0) \rangle = 0, \ \forall \ \xi \in \partial f(x_0), \ \zeta_j \in \partial g_j(x_0). \tag{1}$$

$$\sum_{j=1}^m \nu_j \ g_j(x_0) = 0. \tag{2}$$

For $y \in A$, let us denote by $J_0(y)$ the set of active constraints at y, that is $J_0(y) = \{j : g_j(y) = 0\}$.

Motivated by the work of Caristi, Ferrara and Stefanescu [1], we introduce the following concepts.

Let X be a nonempty set in $\mathbb{R}^n$, f, g_j are nondifferentiable functions and X_0 be a fixed subset of X.

Definition 2 *The pair (f, g_j) for $j = 1, 2, ..., m$, is said to be η-inf-type-I at $x_0 \in X$, with respect to X_0, if*

$$\inf_{x \in X_0} [f(x) - f(x_0)] \geq \inf_{x \in X_0} \langle \xi, \eta(x, x_0) \rangle, \ \forall \ \xi \in \partial f(x_0),$$

$$-g_j(x_0) \geq \inf_{x \in X_0} \langle \zeta_j, \eta(x, x_0) \rangle, \ \forall \ \zeta_j \in \partial g_j(x_0).$$

If in the above definition, the first inequality is strict inequality, then we say that (f, g_j) is η-inf-semistrictly-type-I at $x_0 \in X$, with respect to X_0.

Definition 3 *The pair (f, g_j) for $j = 1, 2, ..., m$, is said to be η-inf-sup-type-I at $x_0 \in X$, with respect to X_0, if*

$$\inf_{x \in X_0} [f(x) - f(x_0)] \geq \inf_{x \in X_0} \langle \xi, \eta(x, x_0) \rangle, \ \forall \ \xi \in \partial f(x_0),$$

$$-g_j(x_0) \geq \sup_{x \in X_0} \langle \zeta_j, \eta(x, x_0) \rangle, \ \forall \ \zeta_j \in \partial g_j(x_0).$$

Definition 4 *The pair (f, g_j) for $j = 1, 2, ..., m$, is said to be η-sup-quasi-type-I at $x_0 \in X$, with respect to X_0, if*

$$\sup_{x \in X_0} [f(x) - f(x_0)] \leq 0 \Rightarrow \sup_{x \in X_0} \langle \xi, \eta(x, x_0) \rangle \leq 0, \ \forall \ \xi \in \partial f(x_0),$$

$$-g_j(x_0) \leq 0 \Rightarrow \sup_{x \in X_0} \langle \zeta_j, \eta(x, x_0) \rangle \leq 0, \ \forall \ \zeta_j \in \partial g_j(x_0).$$

Definition 5 *The pair (f, g_j) for $j = 1, 2, ..., m$, is said to be η-inf-pseudo-type-I at $x_0 \in X$, with respect to X_0, if*

$$\inf_{x \in X_0} \langle \xi, \eta(x, x_0) \rangle \geq 0 \Rightarrow \inf_{x \in X_0} [f(x) - f(x_0)] \geq 0, \ \forall \ \xi \in \partial f(x_0),$$

$$\inf_{x \in X_0} \langle \zeta_j, \eta(x, x_0) \rangle \geq 0 \Rightarrow -g_j(x_0) \geq 0, \ \forall \ \zeta_j \in \partial g_j(x_0).$$

Definition 6 *The pair (f, g_j) for $j = 1, 2, ..., m$, is said to be η-sup-quasi-inf-pseudo-type-I at $x_0 \in X$, with respect to X_0, if*

$$\sup_{x \in X_0} [f(x) - f(x_0)] \leq 0 \Rightarrow \sup_{x \in X_0} \langle \xi, \eta(x, x_0) \rangle \leq 0, \ \forall \ \xi \in \partial f(x_0),$$

$$\inf_{x \in X_0} \langle \zeta_j, \eta(x, x_0) \rangle \geq 0 \Rightarrow -g_j(x_0) \geq 0, \ \forall \ \zeta_j \in \partial g_j(x_0).$$

If in the above definition, the second inequality is satisfied as

$$\inf_{x \in X_0} \langle \zeta_j, \eta(x, x_0) \rangle \geq 0 \Rightarrow -g_j(x_0) > 0, \ \forall \ \zeta_j \in \partial g_j(x_0)$$

then we say that (f, g_j) η-sup-quasi-inf-strictly pseudo-type-I at $x_0 \in X$, with respect to X_0.

Definition 7 *The pair (f, g_j) for $j = 1, 2, ..., m$, is said to be η-inf-pseudo-sup-quasi-type-I at $x_0 \in X$, with respect to X_0, if*

$$\inf_{x \in X_0} \langle \xi, \eta(x, x_0) \rangle \geq 0 \Rightarrow \inf_{x \in X_0} [f(x) - f(x_0)] \geq 0, \ \forall \ \xi \in \partial f(x_0),$$

$$-g_j(x_0) \leq 0 \Rightarrow \sup_{x \in X_0} \langle \zeta_j, \eta(x, x_0) \rangle \leq 0, \ \forall \ \zeta_j \in \partial g_j(x_0).$$

If in the above definition, the second inequality is satisfied as

$$\inf_{x \in X_0} \langle \xi, \eta(x, x_0) \rangle \geq 0 \Rightarrow \inf_{x \in X_0} [f(x) - f(x_0)] > 0, \ \forall \ \xi \in \partial f(x_0)$$

then we say that (f, g_j) η-inf-strictly pseudo-sup-quasi-type-I at $x_0 \in X$, with respect to X_0.

3   Necessary and sufficient optimality conditions

In this section we establish necessary and sufficient optimality condition for nonsmooth constrained optimization problems.

Martin [7] showed that if the constraints of a mathematical programming problem are linear and the feasible set is bounded, then the objective function must be convex. Because of this, he introduced a weaker invexity-type concept which is necessary and sufficient for the sufficiency of the Kuhn-Tucker conditions. Later Caristi, Ferrara and Stefanescu [1] extended the concept of Martin [7] to the setting of η-inf-sup-invexity. To obtain the necessary optimality condition, we generalize Theorem 5 in [1]. The proof of the following theorem follows along the similer lines to the proof of Theorem 5 in [1], so we omitt the proof.

Theorem 1 *(Necessary Optimality Condition) Every Kuhn-Tucker point of problem (P) is a global minimizer if and only if there exists a vector function $\eta : X \times X \rightarrow \mathbb{R}^n$, such that (f, g_j) is η-inf-pseudo-sup-quasi-type-I on X, with respect to A, whenever $j \in J_0(y)$.*

Theorem 2 *(Sufficient Optimality Condition) Let (x_0, ν) be a Kuhn-Tucker point of the problem (P). If there exists η such that (f, g_j) is η-inf-pseudo-sup-quasi-type-I at $x_0 \in X$, with respect to A, for $j \in J_0(x_0)$, then x_0 is an optimal solution of (P).*

Proof. From (1) and (2) it follows that

$$\inf_{x\in A} \langle \xi, \eta(x,x_0)\rangle + \inf_{x\in A} \sum_{j\in J_0(x_0)} \nu_j \langle \zeta_j, \eta(x,x_0)\rangle = 0,\ \forall\, \xi \in \partial f(x_0),\ \zeta_j \in \partial g_j(x_0)$$

$$\Rightarrow \inf_{x\in A}\langle \xi,\eta(x,x_0)\rangle = -\inf_{x\in A}\sum_{j\in J_0(x_0)}\nu_j\langle\zeta_j,\eta(x,x_0)\rangle \geq \sum_{j\in J_0(x_0)} -\nu_j\sup_{x\in A}\langle\zeta_j,\eta(x,x_0)\rangle.$$

Since $g_j(x_0)=0$ whenever $j\in J_0(x_0)$, it follows from the η-inf-pseudo part of the assumption that

$$\sum_{j\in J_0(x_0)}(-\nu_j)\sup_{x\in A}\langle\zeta_j,\eta(x,x_0)\rangle \geq 0,\ \forall\ \zeta_j\in\partial g_j(x_0)$$

and therefore

$$\inf_{x\in A}\langle\xi,\eta(x,x_0)\rangle \geq 0,\ \forall\,\xi\in\partial f(x_0).$$

The η-inf-pseudo part of the assumption implies that

$$\inf_{x\in A}[f(x)-f(x_0)]\geq 0.$$

Therefore x_0 is an optimal solution of (P).

Corollary 1 *Let (x_0,ν) be a Kuhn-Tucker point of the problem (P). If there exists η such that (f,g_j) is η-inf-sup-type-I at $x_0\in X$, with respect to A, for $j\in J_0(x_0)$, then x_0 is an optimal solution of (P).*

4 Wolfe type duality

Now in relation to (P) we consider the following Wolfe type dual problem given as in Caristi, Ferrara and Stefanescu [1]:

$$\textbf{(D)}\quad \sup_{(x,\nu)\in V}\ (f(x)+\sum_{j=1}^m \nu_j\, g_j(x)),$$

where

$$V=\left\{(x,\nu)\in X\times R_+^m : \langle\xi,\eta(x,y)\rangle + \sum_{j=1}^m \nu_j\langle\zeta_j,\eta(x,y)\rangle = 0,\ \forall\,\xi\in\partial f(x),\ \zeta_j\in\partial g_j(x)\right\}.$$

Theorem 3 *Assume that there exist η such that (f,g_j) is η-inf-sup-type-I on X. Then, for any feasible soluion x of (P) $(x\in A)$, and for any feasible solution (y,ν) of (D) $((y,\nu)\in V)$*

$$f(x)-f(y)\geq \sum_{j=1}^m \nu_j\, g_j(y).$$

Proof. From the hypothesis of the theorem it follows that $\forall\ \xi \in \partial f(y),\ \zeta_j \in \partial g_j(y)$

$$\inf_{x \in A} \langle \xi, \eta(x,y) \rangle - \sum_{j=1}^{m} \nu_j\ g_j(y) = \inf_{x \in A} \sum_{j=1}^{m} -\nu_j\ \langle \zeta_j, \eta(x,y) \rangle - \sum_{j=1}^{m} \nu_j\ g_j(y)$$

$$\geq -\sup_{x \in A} \sum_{j=1}^{m} \nu_j\ \langle \zeta_j, \eta(x,y) \rangle - \sum_{j=1}^{m} \nu_j\ g_j(y)$$

$$\geq \sum_{j=1}^{m} \nu_j\ g_j(y) - \sum_{j=1}^{m} \nu_j\ g_j(y) = 0.$$

Therefore

$$\inf_{x \in A} \left(f(x) - f(y) - \sum_{j=1}^{m} \nu_j\ g_j(y) \right) \geq 0,$$

which proves the theorem.

Now we give a brief extension of the earlier work of Mishra and Rueda [8] and Xu[15] to this class of functions and hence, some results of Xu [9].

Xu [15] introduced a new class of nonlinear programming, called SFJ-invex programming, which lies exactly between invex programming and type-I programming. Following Xu [15], problem (P) is said to be SFJ-invex, if there exist a function $\eta : X_0 \times A \to R^n$, such that

$$(x,u) \in (X_0, A) \Rightarrow \left\{ \begin{array}{l} f(x) - f(u) - \nabla f(u)^T \eta(x,u) \geq 0, \\ \text{if } g_j(u) = 0,\ j = 1, 2, ..., m \text{ then } g_j(x) - \nabla g_j(u)^T \eta(x,u) \geq 0. \end{array} \right\}$$

For the support of this new class of programming one can see Example 2.1 and 2.2 in Xu [15].

Now, we propose the following problem:

Problem (P) is said to be η-inf-SFJ-type-I if there exist a function $\eta : X_0 \times A \to R^n$, for all $\xi \in \partial f(u)$ and $\zeta_j \in \partial g_j(u)$ such that

$$(x,u) \in (X_0, A) \Rightarrow \left\{ \begin{array}{l} \inf_{x \in X_0} [f(x) - f(u)] \geq \inf_{x \in X_0} \langle \xi, \eta(x,u) \rangle, \\ \text{if } g_j(u) = 0,\ j = 1, 2, ..., m \text{ then } \inf_{x \in X_0} g_j(x) \geq \inf_{x \in X_0} \langle \zeta_j, \eta(x,u) \rangle. \end{array} \right\}$$

With this new concept, the results of Xu [15] can be extended to this class of functions. Moreover, we can extend the earlier work of Mishra and Rueda [8] to this new setup.

References

[1] G. Caristi, M. Ferrara and A. Stefanescu, New invexity type conditions in constrained optimization, Proceedings of the 6th International Symposium on Generalized Convexity / Monotonicity, Lecture Notes in Economics and Mathematical Systems No. 502 (ed. by N. Hadjisavvas, J. E. Martinez Legaz, and J. P. Penot), Springer (2001) 159–166.

[2] F.H. Clarke, Optimization and Nonsmooth Analysis. Wiley, New York, 1983.

[3] B.D. Craven and B.M. Glover, Invex functions and duality, Journal of the Australian Mathematical Society Series A 39, 1 (1985) 1–20.

[4] M.A. Hanson, On sufficiency of the KuhnTucker conditions, Journal of Mathematical Analysis and Applications 80, 2 (1981) 545–550.

[5] M.A. Hanson and B. Mond, Necessary and sufficient conditions in constrained optimization, Mathematical Programming 37, 1 (1987) 51–58.

[6] R.N. Kaul, S.K. Suneja and M.K. Srivastava, Optimality criteria and duality in multiple-objective optimization involving generalized invexity, Journal of Optimization Theory and Applications 80, 3 (1994) 465–482.

[7] D.H. Martin, The essence of invexity, Journal of Optimization Theory and Applications 47, 1 (1985) 65–76.

[8] S.K. Mishra and N.G. Rueda, On univexitytype nonlinear programming problems, Bulletin of the Allahabad Mathematical Society 16 (2001) 105–113.

[9] S.K. Mishra and N.G. Rueda, Higher-order generalized invexity and duality in nondifferentiable mathematical programming, Journal of Mathematical Analysis and Applications 272 (2002) 496–506.

[10] S.K. Mishra, S.Y. Wang and K.K. Lai, Optimality and duality in nondifferentiable and multiobjective programming under generalized d-invexity, Journal of Global Optimization 29 (2004) 425-438.

[11] S.K. Mishra, S.Y. Wang and K.K. Lai, Optimality and duality for multiple objective optimization under generalized type-I univexity, Journal of Mathematical Analysis and Applications 303 (2005) 315-326.

[12] S.K. Mishra, S.Y. Wang and K.K. Lai, Multiple objective fractional programming involving semilocally type-I preinvex and related functions, Journal of Mathematical Analysis and Applications 310 (2005) 626-640.

[13] S.K. Mishra, S.Y. Wang and K.K. Lai, Nondifferentiable multiobjective programming under generalized d-univexity, European Journal of Operational Research 160 (2005) 218–226.

[14] N.G. Rueda and M.A. Hanson, Optimality criteria in mathematical programming involving generalized invexity, Journal of Mathematical Analysis and Applications 130, 2 (1988) 375–385.

[15] Z.K. Xu, On invexity type nonlinear programming problems, Journal of Optimization Theory and Applications 80, 1 (1994) 135–148.

CHAPTER 7

Duality theory for the multiobjective nonlinear programming involving generalized convex functions*

R. Osuna-Gomez[†]   M. B. Hernández-Jiménez[‡]   L. L. Salles Neto [§]

Abstract

In this chapter we present and prove different forms of weak, strong and converse duality theorems for the Wolfe and Mond-Weir dual problems associated with the vector optimization problem with constraints, where the vector objective function and the vector function associated with the inequality-type constraints are invex, strictly invex or quasi-invex.

Keywords: Multiobjective programming, invexity, duality, optimality conditions, efficient solutions.

1   Introduction

Another approach to solving mathematical programming problems is through the theory of duality. The basic idea is to associate the original problem, called primal, with another problem, called dual, such that the dual problem provides bounds for the optimal values of primal and under certain conditions by solving one also solve the other. In the latest decade several papers have been published aiming to characterize and relate the solutions of a vector optimization problem with the solutions of dual associated problems [6], [1], [9], [5], [8], [4], [10], but there are few papers that involve weakly efficient solutions. In addition, working with these solutions offers the advantage that the formulations of the dual problems are more general than those found in the literature.

Let us consider the following vector optimization problem (VOP)

$$\textbf{(VOP)} \quad \begin{aligned} Min \quad & f(x), \\ s.t.: \quad & g(x) \leqq 0, \\ & x \in S \end{aligned}$$

---

*This work was partially supported by the grant MTM2007-063432 of the Science and Education Spanish Ministry. The third author also was supported in part by the FAPESP grant no. 2009/06139-5.

[†]Departamento de Estadística e Investigación Operativa, Facultad de Matemáticas, Universidad de Sevilla, Spain, e-mail: rafaela@us.es

[‡]Departamento de Economía, Métodos Cuantitativos e H^a Económica. Area de Estadística e Investigación Operativa. Universidad Pablo de Olavide, Spain, e-mail: mbherjim@upo.es

[§]Departamento de Ciência e Tecnologia, Universidade Federal de São Paulo, Brazil, e-mail: luiz.leduino@unifesp.br

where S is an open subset of $\mathbb{R}^n$, $f = (f_1, \ldots, f_p) : S \subseteq \mathbb{R}^n \to \mathbb{R}^p$ and $g = (g_1, \ldots, g_m) : S \subseteq \mathbb{R}^n \to \mathbb{R}^m$ are differentiable.

For a precise definition of the Pareto-optimal solution (efficient) for (VOP) [7], the following convention for equalities and inequalities is assumed below:

If $x = (x_1, \ldots, x_n)$, $y = (y_1, \ldots, y_n) \in \mathbb{R}^n$, then

$$
\begin{aligned}
x = y &\Leftrightarrow x_i = y_i, \quad \forall i = 1, \ldots, n, \\
x < y &\Leftrightarrow x_i < y_i, \quad \forall i = 1, \ldots, n, \\
x \leqq y &\Leftrightarrow x_i \leq y_i, \quad \forall i = 1, \ldots, n, \\
x \leq y &\Leftrightarrow x_i \leq y_i, \quad \forall i = 1, \ldots, n, \quad \text{and there exist } j \text{ such that } x_j < y_j.
\end{aligned}
$$

Similarly, $>, \geqq, \geq$.

We can now define efficient solution and weakly efficient solution for (VOP)

Definition 1 *A feasible point, $\bar{x}$, is said to be an efficient solution for (VOP) if there does not exist another feasible point, x, such that $f(x) \leq f(\bar{x})$.*

It was later defined a more general concept: weakly efficient solution for (VOP).

Definition 2 *A feasible point, $\bar{x}$, is said to be a weakly efficient solution for (VOP) if there does not exist another feasible point, x, such that $f(x) < f(\bar{x})$.*

It is easy to prove that any efficient solution is a weakly efficient solution.

The chapter is organized as follows. In the following section we define the Wolfe Dual and Mond-Weir Dual for the vector optimization problem and present some basic theorems involved. The next section reports the weak, strong and converse duality theorems to the Wolfe Dual, and in the following section we present the equivalent theorems to the Mond-Weir Dual. The next section presents more powerful converse duality theorem under the assumption that the functions f and g are twice differentiable. Finally, we state the conclusions.

2 Duality theory for the vector optimization problem

Let X be the feasible set of the primal problem (VOP):

$$X = \{x \in S \subseteq \mathbb{R}^n, g(x) \leqq 0\}$$

Definition 3 *A feasible point $\bar{x}$ for (VOP) is said to be a Kuhn-Tucker vector critical point, (hereafter KTVCP), if there exists $\lambda \in \mathbb{R}^p$, $\bar{y} \in \mathbb{R}^m$ such that*

$$\lambda^T \nabla f(\bar{x}) + \bar{y}^T \nabla g(\bar{x}) = 0 \tag{1}$$

$$\bar{y}^T g(\bar{x}) = 0 \tag{2}$$

$$\bar{y} \geqq 0 \tag{3}$$

$$\lambda \geq 0 \tag{4}$$

The following result is due to Craven [2], which is basic to proof the duality theorems related to the dual problem.

Theorem 1 *Let $\bar{x}$ be a weakly efficient solution for (VOP), and assume that $\bar{x}$ satisfies a constraint qualification. Then there exists $\bar{\lambda} \in \mathbb{R}^p$ with $\bar{\lambda} \geq 0$ and $\bar{y} \in \mathbb{R}^m$ with $\bar{y} \geqq 0$ such that:*

$$\bar{\lambda}^T \nabla f(\bar{x}) + \bar{y}^T \nabla g(\bar{x}) = 0,$$

$$\bar{y}^T g(\bar{x}) = 0.$$

Definition 4 *The Wolfe Multiobjective Dual problem associated with (VOP) is formulated as:*

$$
\begin{aligned}
\textbf{(DW)} \quad Max \quad & f(u) + \bar{y}^T g(u)e, \\
s.t.: \quad & \lambda^T \nabla f(u) + \bar{y}^T \nabla g(u) = 0, \\
& \bar{y} \geqq 0, \\
& \lambda \geq 0, \quad \lambda^T e = 1, \\
& u \in S.
\end{aligned}
$$

where λ and $e = (1, 1, ..., 1)^T \in \mathbb{R}^p$.

Definition 5 *The Mond-Weir Multiobjective Dual problem associated with (VOP) is formulated as:*

$$
\begin{aligned}
\textbf{(DMW)} \quad Max \quad & f(u), \\
s.t.: \quad & \lambda^T \nabla f(u) + \bar{y}^T \nabla g(u) = 0, \\
& \bar{y}^T g(u) \geq 0, \\
& \bar{y} \geqq 0, \\
& \lambda > 0, \quad \lambda^T e = 1, \\
& u \in S.
\end{aligned}
$$

where λ and $e = (1, 1, ..., 1)^T \in \mathbb{R}^p$.

Remark 1 *The dual problems which can be found in the literature require that the vector λ be strictly positive.*

We note by U_W and U_{MW} the feasible sets for Wolfe and Mond-Weir Dual problems respectively.

In the next sections we need the following definitions:

Definition 6 *Let $f = (f_1, \ldots, f_p) : S \subseteq \mathbb{R}^n \to \mathbb{R}^p$ be a differentiable function defined on the open set S. Then:*

a) the vector function f is said to be invex on S if there exist a vector function $\eta :$ $\mathbb{R}^n \times \mathbb{R}^n \to \mathbb{R}^n$ such that $\forall x, \bar{x} \in S$

$$f(x) - f(\bar{x}) \geqq \eta(x, \bar{x})^T \nabla f(\bar{x})$$

b) the vector function f is said to be quasi-invex on S if there exist a vector function $\eta : \mathbb{R}^n \times \mathbb{R}^n \to \mathbb{R}^n$ such that $\forall x, \bar{x} \in S$

$$f(x) - f(\bar{x}) \leqq 0 \Rightarrow \eta(x, \bar{x})^T \nabla f(\bar{x}) \leqq 0$$

3 Wolfe type duality

Here we show that the hypothesis in previous theorems can be generalized to the class of invex and quasi-invex functions that also include strictly pseudoconvex and quasiconvex functions.

We establish the corresponding results considering the Wolfe dual problem and weakly efficient solutions for multiobjective problems.

Theorem 2 (Weak duality):

Let $\bar{x} \in X$ and $(\bar{u}, \bar{\lambda}, \bar{y}) \in U_W$ be any feasible solutions. Let f and g be invex functions on S with respect to the same function η. Then it cannot occur that:

$$f(\bar{x}) < f(\bar{u}) + \bar{y}^T g(\bar{u})e.$$

Proof:

Let us reason by contradiction.

Suppose that $\exists \bar{x} \in X$, $\exists (\bar{u}, \bar{\lambda}, \bar{y}) \in U_W$ such that:

$$f(\bar{x}) < f(\bar{u}) + \bar{y}^T g(\bar{u})e$$

then

$$\bar{\lambda}^T f(\bar{x}) < \bar{\lambda}^T f(\bar{u}) + \bar{\lambda}^T \bar{y}^T g(\bar{u})e = \bar{\lambda}^T f(\bar{u}) + \bar{y}^T g(\bar{u})$$

Since $\bar{y} \geqq 0$ y $g(\bar{x}) \leqq 0 \Longrightarrow \bar{y}^T g(\bar{x}) \leq 0$

Then:

$$\bar{\lambda}^T f(\bar{x}) + \bar{y}^T g(\bar{x}) < \bar{\lambda}^T f(\bar{u}) + \bar{y}^T g(\bar{u})$$

$$0 > (\bar{\lambda}^T f(\bar{x}) + \bar{y}^T g(\bar{x})) - (\bar{\lambda}^T f(\bar{u}) + \bar{y}^T g(\bar{u})) \geq$$

$$\geq \eta(\bar{x}, \bar{u})^T (\bar{\lambda}^T \nabla f(\bar{u}) + \bar{y}^T \nabla g(\bar{u})) = 0$$

But that is an absurd.   □

Theorem 3 (Strong Duality):

Let x^* be a weakly efficient solution for (VOP), and assume that x^* satisfies a constraint qualification. Then there exists (λ^*, y^*) such that (x^*, λ^*, y^*) is feasible for (DW) and the values of the objective function of (VOP) and (DW) are equal. Moreover, if f and g are invex with respect to the same vector function η on S then (x^*, λ^*, y^*) is a weakly efficient solution for (DW).

Proof:

If x^* is a weakly efficient solution for (VOP), by Theorem 1, there exists $\bar{\lambda} \geq 0$ and $\bar{y} \geqq 0$ such that (x^*, λ^*, y^*) is a KTVCP for the (VOP), i.e.

$$\lambda^{*T}\nabla f(x^*) + y^{*T}\nabla g(x^*) = 0 \quad (1)$$
$$\bar{y}^T g(x^*) = 0 \quad (2)$$

By (1), $(x^*, \lambda^*, y^*) \in U_W$ and by (2) the values of the objective functions of (VOP) and (DW) are equal.

If in addition f and g are invex with respect to η there can be no other $(u, \lambda, y) \in U_W$ such that,

$$f_i(u) + y^T g(u) > f_i(x^*) + y^{*T} g(x^*) = f_i(x^*), \quad \forall i$$

because it would contradict the weak duality theorem.

$\square$

For the next theorem, we need to establish a stronger invex assumption on the objective function.

Definition 7 *We say that a p-dimensional function f is strictly invex on S with respect to the function η if*

$$\forall x_1, x_2 \in S, \ x_1 \neq x_2, \quad \exists \eta(x_1, x_2) \subset \mathbb{R}^n$$

such that

$$f(x_1) - f(x_2) > \eta(x_1, x_2)^T \nabla f(x_2)$$

Theorem 4 (Converse Duality):

Let $\bar{x} \in X$, $(\bar{u}, \bar{\lambda}, \bar{y}) \in U_W$ be such that $\bar{\lambda}^T f(\bar{x}) = \bar{\lambda}^T f(\bar{u}) + \bar{y}^T g(\bar{u})$. If f is strictly invex and g is invex with respect to the same vector function η on S then $\bar{x} = \bar{u}$.

Proof:

Suppose that $\bar{x} \neq \bar{u}$.

If g is invex with respect to η then there also exists a non-negative linear combination $\bar{y}^T g$ with respect to the same vector.

On the other hand we have $\bar{y}^T g(\bar{x}) \leq 0$ since $\bar{y} \geqq 0$ and $g(\bar{x}) \leqq 0$.

Therefore:

$$-\bar{y}^T g(\bar{u}) \geq \bar{y}^T g(\bar{x}) - \bar{y}^T g(\bar{u}) \geq \eta(\bar{x}, \bar{u})^T \bar{y}^T \nabla g(\bar{u}) =$$
$$= -\eta(\bar{x}, \bar{u})^T \bar{\lambda}^T \nabla f(\bar{u})$$

So,

$$-\bar{y}^T g(\bar{u}) \geq -\eta(\bar{x}, \bar{u}) \bar{\lambda}^T \nabla f(\bar{u}) \iff$$
$$\iff \bar{y}^T g(\bar{u}) \leq \eta(\bar{x}, \bar{u})^T \bar{\lambda}^T \nabla f(\bar{u})$$

For f be strictly invex,

$$f(\bar{x}) - f(\bar{u}) > \eta(\bar{x}, \bar{u})^T \nabla f(\bar{u})$$

Multiplying by $\bar{\lambda}$

$$\bar{\lambda}^T f(\bar{x}) - \bar{\lambda}^T f(\bar{u}) > \eta(\bar{x}, \bar{u})^T \bar{\lambda}^T \nabla f(\bar{u}) \geq \bar{y}^T g(\bar{u})$$

And so

$$\bar{\lambda}^T f(\bar{x}) > \bar{\lambda}^T f(\bar{u}) + \bar{y}^T g(\bar{u})$$

which contradicts the hypothesis.

□

Another converse duality theorem can be stated as follows,

Theorem 5 (Converse Duality):

Let $(\bar{u}, \bar{\lambda}, \bar{y}) \in U_W$ be such that $\bar{u} \in X$. If f and g are invex with respect to the same vector function η, then $\bar{u}$ is a weakly efficient solution for (VOP).

Proof:
Suppose that $\exists \hat{x}$ such that $g(\hat{x}) \leqq 0$ and

$$f(\hat{x}) < f(\bar{u}) \leq f(\bar{u}) + \bar{y}^T g(\bar{u})$$

which contradicts the weak duality Theorem 2.

□

In previous demonstrations we concluded that these theorems remain valid even if it would change the invex hypothesis by other more general ones.

Theorem 6 (Weak Duality):
If $x \in X$ and $(u, \lambda, \bar{y}) \in U_W$ and if one of the following assumptions is satisfied:

a) $f + \bar{y}^T g$ is invex on S, or

b) $\lambda^T f + \bar{y}^T g$ is invex on S

then it cannot occur that:

$$f(x) < f(u) + \bar{y}^T g(u)e.$$

And from the weak duality theorem we can show the following strong duality result.

Theorem 7 (Strong Duality):

If $\bar{x}$ is a weakly efficient solution for (VOP) and $\bar{x}$ satisfies a constraint qualification then there exists $(\bar{\lambda}, \bar{y}) \geq 0$ such that $(\bar{x}, \bar{\lambda}, \bar{y}) \in U_W$ and the values of the objective functions of (VOP) and (DW) are equal.

If in addition:

a) $f + \bar{y}^T g$ is invex on S or

b) $\lambda^T f + \bar{y}^T g$ is invex on S

then $(\bar{x}, \bar{\lambda}, \bar{y})$ is a weakly efficient solution for (DW).

4   Mond Weir type duality

For the Mond-Weir dual multiobjective problem we can prove similar results under invexity hypothesis. We continue considering the weakly efficient solutions.

Theorem 8 (Weak Duality):

Let $x \in X$ and $(u, \lambda, \bar{y}) \in U_{MW}$. If f and g are invex with respect to the same vector function η, then it cannot occur that:

$$f(x) < f(u).$$

Proof:
Suppose that there exist feasible points $x \in X$, and $(u, \lambda, \bar{y}) \in U_{MW}$ such that

$$f(x) < f(u) \implies \lambda^T f(x) < \lambda^T f(u)$$

If $x \in X \implies \bar{y}^T g(x) \leq 0$ then,

$$\lambda^T f(x) + \bar{y}^T g(x) < \lambda^T f(u) + \bar{y}^T g(u)$$

$$0 > (\lambda^T f(x) + \bar{y}^T g(x)) - (\lambda^T f(u) + \bar{y}^T g(u)) \geq \qquad (5)$$

$$\geq \eta(x,u)^T(\lambda^T \nabla f(u) + \bar{y}^T \nabla g(u))$$

since f and g are invex with respect to the η.

But $(u, \lambda, \bar{y}) \in U_M$

$$\eta(x,u)^T(\lambda^T \nabla f(u) + \bar{y}^T \nabla g(u)) = 0$$

which is contradictory to (5).

$\square$

The strong duality theorem is immediate from the previous result.

Theorem 9 (Strong Duality):

Let $\bar{x}$ be a weakly efficient solution for (VOP), and assume that $\bar{x}$ satisfies a constraint qualification. If f and g are invex with respect to the same vector function η, then there exists $\bar{\lambda}, \bar{y}$ such that $(\bar{x}, \bar{\lambda}, \bar{y})$ is a weakly efficient solution for (DMW).

Proof:

If $\bar{x}$ is a weakly efficient solution for (VOP) then there exists $\bar{\lambda}, \bar{y}$ such that $(\bar{x}, \bar{\lambda}, \bar{y})$ is a KTVCP and therefore $(\bar{x}, \bar{\lambda}, \bar{y}) \in U_{MW}$.

$\square$

Theorem 10 (Converse Duality):

Let $\bar{x} \in X$ and $(\bar{u}, \bar{\lambda}, \bar{y}) \in U_{MW}$ such that $\bar{\lambda}^T f(\bar{x}) = \bar{\lambda}^T f(\bar{u})$ with f strictly invex and g invex on S with respect to the same vector function . Then $\bar{x} = \bar{u}$

Proof:

Suppose that $\bar{x}$ and $\bar{u}$ are different.
Using that g is invex and $(\bar{u}, \bar{\lambda}, \bar{y}) \in U_M$:

$$0 \geq \bar{y}^T g(\bar{x}) - \bar{y}^T g(\bar{u}) \geq -\eta(\bar{x},\bar{u})^T \bar{\lambda}^T \nabla f(\bar{u}) \Longrightarrow$$

$$\Longrightarrow \eta(\bar{x},\bar{u})^T \bar{\lambda}^T \nabla f(\bar{u}) \geq 0$$

As f is strictly invex

$$\bar{\lambda}^T f(\bar{x}) - \bar{\lambda}^T f(\bar{u}) > \bar{\lambda}^T \eta(\bar{x},\bar{u})^T \nabla f(\bar{u}) \geq 0$$

Then

$$\bar{\lambda}^T f(\bar{x}) > \bar{\lambda}^T f(\bar{u})$$

which contradicts the hypothesis.

$\square$

We can also state the converse duality theorem as follows:

Theorem 11 (Converse Duality):

Let $(\bar{u}, \bar{\lambda}, \bar{y}) \in U_{MW}$ such that $\bar{u}$ is a feasible point for the primal problem. If f and g are invex with respect to the same vector function η then $\bar{u}$ is a weakly efficient solution for (VOP).

Proof:

If $(\bar{u}, \bar{\lambda}, \bar{y}) \in U_M$ and $\bar{u} \in X$ we have:

$$\bar{\lambda}^T \nabla f(\bar{u}) + \bar{y}^T \nabla g(\bar{u}) = 0$$
$$\bar{y}^T g(\bar{u}) = 0$$

So $\bar{u}$ is a KTVCP for (VOP) and with the assumption of invexity of the functions f and g we have that $\bar{u}$ is a weakly efficient point for the primal problem.

$\square$

The above theorems can be proved under slightly different hypothesis, such as

Theorem 12 (Weak Duality):

Let $x \in X$ and $(u, \lambda, \bar{y}) \in U_{MW}$. If

a) f is invex and $\bar{y}^T g$ is quasi-invex with respect to the same η or

b) f is quasi-invex and $\bar{y}^T g$ is strictly invex with respect to the same η or

c) $\lambda^T f$ is quasi-invex and $\bar{y}^T g$ is strictly invex with respect to the same η

then it cannot occur that:

$$f(x) < f(u).$$

Proof:

a) Suppose that $f_i(x) < f_i(u) \; \forall i$

$$\Longrightarrow \eta(x, u)^T \nabla f_i(u) < 0 \quad \forall i$$

Since $\lambda \geq 0 \Longrightarrow \lambda^T \eta(x, u)^T \nabla f(u) < 0$.
On the other hand

$$y^T g(x) - y^T g(u) \leq 0 \Longrightarrow \eta(x, u)^T y^T \nabla g(u) \leq 0$$

Adding

$$\eta(x, u)^T (\lambda^T \nabla f(u) + y^T \nabla g(u)) < 0$$

and it is a contradiction with the dual feasibility of (u, λ, y).

b) By similar reasoning we obtain:

$$\lambda^T \eta(x,u)^T \nabla f(u) \le 0$$

$$\eta(x,u)^T y^T g(u) < 0$$

Then

$$\eta(x,u)^T (\lambda^T \nabla f(u) + y^T g(u)) < 0$$

and it is a contradiction with the feasibility of x.

c) It follows from the same argument in the preceding paragraph.

$\square$

With the above assumptions we can also prove strong and converse duality theorems such as the following:

Theorem 13 (Strong Duality:)

Let $\bar{x}$ be a weakly efficient solution for (VOP) and assume that $\bar{x}$ satisfies a constraint qualification, then $\exists\,(\bar\lambda,\bar y)$ such that $(\bar x,\bar\lambda,\bar y)\in U_{MW}$ and the values of objective functions are the same.

Furthermore, if any of the conditions a), b) or c) of the previous theorem is satisfied then $(\bar x,\bar\lambda,\bar y)$ is a weakly efficient solution for (DMW).

Theorem 14 (Strict Converse Duality):

Let $\bar{x}$ be a weakly efficient solution for (VOP) and $(\bar u,\bar\lambda,\bar y)$ be a weakly efficient solution for (DMW) such that $\bar\lambda^T f(\bar x) \le \bar\lambda^T f(\bar u)$.

Suppose that

1. $\bar\lambda^T f$ is strictly invex and $\bar y^T g$ is quasi-invex with respect to the same η or

2. $\bar\lambda^T f$ is quasi-invex and $\bar y^T g$ is strictly invex at $\bar y$ with respect to the same η

Then $\bar x = \bar u$.

Proof:

Suppose that $\bar x \ne \bar u$

1. Since $\bar x \in X$ and $(\bar x,\bar\lambda,\bar y) \in U_M \Longrightarrow$

$$\bar y^T g(\bar x) - \bar y^T g(\bar u) \le 0 \Longrightarrow$$

$$\Longrightarrow \eta(\bar x,\bar u)^T \bar y^T \nabla g(\bar u) \le 0 \Longrightarrow$$

$$\Longrightarrow \eta(\bar x,\bar u)^T \bar\lambda^T \nabla f(\bar u) \ge 0 \Longrightarrow$$

$$\Longrightarrow \bar\lambda^T f(\bar x) > \bar\lambda^T f(\bar u)$$

which contradicts the hypothesis.

2. Using the assumption of strict invexity,

$$\bar{y}^T g(\bar{x}) - \bar{y}^T g(\bar{u}) \leq 0 \Longrightarrow$$

$$\Longrightarrow \eta(\bar{x}, \bar{u})^T \bar{y}^T \nabla g(\bar{u}) < 0$$

$$\Longrightarrow \eta(\bar{x}, \bar{u})^T \bar{\lambda}^T \nabla f(\bar{u}) > 0$$

$$\Longrightarrow \bar{\lambda}^T f(\bar{x}) > \bar{\lambda}^T f(\bar{u})$$

which contradicts the hypothesis.

$$\square$$

5 Duality: twice differentiable functions

When f and g are twice differentiable functions we may prove a more powerful converse duality theorem, in the sense that we can find weakly efficient solutions for the primal problem from weakly efficient solutions for dual problems.

Then we state and prove these theorems for the case of Wolfe and Mond-Weir dual problems previously defined.

Theorem 15 (Converse Duality):

Let f and g be invex for the same η and suppose they are twice differentiable on S, and let $(\bar{u}, \bar{\lambda}, \bar{y})$ be a weakly efficient solution for (DW). If the Hessian matrix

$$\bar{\lambda}^T \nabla^2 f(\bar{u}) + \bar{y}^T \nabla^2 g(\bar{u})$$

is a positive or negative definite matrix, then $\bar{u}$ is a weakly efficient solution for (VOP).

Proof:

If $(\bar{u}, \bar{\lambda}, \bar{y})$ is a weakly efficient solution for (DW), then it is a KTVCP for this problem.

By Craven, [3], there exists

$$\tau \in \mathbb{R}^p, v \in \mathbb{R}^m, s \in \mathbb{R}^m, w \in \mathbb{R}^p, q \in \mathbb{R}$$

such that

$$\tau^T \nabla f(\bar{u}) + q\bar{y}^T \nabla g(\bar{u}) + v^T \nabla(\bar{\lambda}^T \nabla f(\bar{u}) + \bar{y}^T \nabla g(\bar{u})) = 0 \tag{6}$$

$$v^T \nabla g(\bar{u}) + qg(\bar{u}) + s = 0 \tag{7}$$

$$v \nabla f(\bar{u})^T + w = 0 \tag{8}$$

$$q\bar{y}^T g(\bar{u}) = 0 \tag{9}$$

$$s^T \bar{y} = 0 \tag{10}$$

$$w^T \bar{\lambda} = 0 \tag{11}$$

$$(\tau, \upsilon, s, w) \geq 0 \tag{12}$$

$$q = \tau^T e \tag{13}$$

Multiplying (11) by e and and using that $\bar{\lambda}^T e = 1 \implies w = 0$.

Thus

$$\upsilon^t \nabla f(\bar{u}) = 0 \tag{14}$$

Multiplying (7) by $\bar{y}^T$

$$\upsilon^T \bar{y}^T \nabla g(\bar{u}) + q\bar{y}^T g(\bar{u}) + \bar{y}^T s = 0$$

Using (9) and (10) $\implies$

$$\upsilon^T \bar{y}^T \nabla g(\bar{u}) = 0 \tag{15}$$

Multiplying (14) by υ

$$(\tau^T \nabla f(\bar{u}) + \upsilon^T \nabla (\bar{\lambda}^T \nabla f(\bar{u}) + \bar{y}^T \nabla g(\bar{u})) + q\bar{y}^T \nabla g(\bar{u}))\upsilon = 0$$

By (14) and (15), the previous equality becomes:

$$\upsilon^T (\bar{\lambda}^T \nabla^2 f(\bar{u}) + \bar{y}^T \nabla^2 g(\bar{u}))\upsilon = 0$$

By hypothesis it can only happen if $\upsilon = 0$

If $\tau = 0 \implies q = 0$. Thus from (7) we have that $s = 0$, which contradicts (13).

Therefore $\tau \geq 0$ and by (13) we have that $q > 0$.

By (7) and using (12) we conclude

$$g(\bar{u}) \leqq 0$$

Therefore by Theorem 5, $\bar{u}$ is a weakly efficient solution for (VOP). $\square$

The previous theorem has its counterpart for the Mond-Weir dual.

Theorem 16 (Converse Duality):

Let f and g be invex for the same η and suppose they are twice differentiable on S, and let $(\bar{u}, \bar{\lambda}, \bar{y})$ be a weakly efficient solution for (DW). If the Hessian matrix

$$\bar{\lambda}^T \nabla^2 f(\bar{u}) + \bar{y}^T \nabla^2 g(\bar{u})$$

is a positive or negative definite matrix and the vectors $\{\nabla f_i(\bar{x})\}_{i=1,\cdots,p}$ are linearly independent, then $\bar{u}$ is a weakly efficient solution for (VOP).

The demonstration follows the same line as that of the previous theorem.

6 Conclusions

In this chapter we have shown interesting theoretical results for the vector optimization problem involving generalized convex functions. We presented weak, strong and converse duality theorems for the Wolfe Dual and Mond-Weir Dual problems. Those results include, under twice differentiability hypothesis of the functions, that we can find weakly efficient solutions for the primal problem from weakly efficient solutions for dual problems. The characterization of weakly efficient solutions is certainly an interesting field for further research in multiobjective nonlinear programming.

References

[1] G. Bitran, Duality in nonlinear multiple criteria optimization problems, J. Optim. Theory Appl. 35-3 (1981) 367-401.

[2] D.B. Craven, Lagrangian Conditions and Quasiduality, B. Austral. Math. Soc. Ser. B 16 (1977) 325-339.

[3] B.D. Craven, Invex Functions and Constraint Local Minimal, J. Austral. Math. Soc. Ser. B 24 (1981) 357-366.

[4] R.R. Egudo, Efficiency and generalized convex duality for multiobjective programs, J. Math. Anal. Appl. 138 (1989) 84-94.

[5] R.R. Egudo and M.A. Hanson, Duality with Generalized Convexity, J. Austral. Math. Soc. Ser. B 28 (1986) 10-21.

[6] B. Mond and T. Weir, Generalized concavity and duality. In: Schaible, S. and Ziemba, W.T. (eds.), Generalized Concavity Optimization and Economics. Academic Press, New York, (1981), 263-280.

[7] V. Pareto, Course d'economie politique, Rouge, Lausanne, 1896.

[8] T. Weir, A note on invex functions and duality in multiple objective optimization, Opsearch 25 (1988) 99-104.

[9] T. Weir, Proper efficiency and duality for vector valued optimization problems, J. Austral. Math. Soc. Ser. A 43 (1987) 21-34.

[10] T. Weir and B. Mond, Generalized convexity and duality in multiobjective programming, Bull. Aust. Math. Soc. 39 (1989) 287-299.

CHAPTER 8

Mixed type duality for multiobjective optimization problems with set constraints

Riccardo Cambini and Laura Carosi*

Abstract

The aim of this chapter is to propose some pairs of dual programs where the primal is a vector problem having a feasible region defined by a set constraint, equality and inequality constraints, while the duals can be classified as "mixed type" ones. The duality results are proved under suitable generalized concavity properties. In this light, the role of different kinds of generalized concavity properties will be deepened on.
Keywords: Vector Optimization, Duality, Maximum Principle Conditions,Generalized Convexity, Set Constraints.
AMS - 2000 Math. Subj. Class.: 90C29, 90C46, 90C26
JEL - 1999 Class. Syst.: C61

1 Introduction

Duality for vector optimization problems has been one of the main issues throughout different fields such as operation research, economic theory, location theory, management science, theory of computational algorithms. Since the pioneer paper by Tanino and Sawaragi [35], duality theory in multiobjective programs has been extensively studied as the huge number of papers and books on this topics witnesses (see for example [3, 16, 22, 27, 34], and references therein). Like in the scalar case, duality has been investigated in different contexts: both smooth and nonsmooth cases have been studied and special attention has been given to particular classes of vector optimization problems such as quadratic problems, generalized fractional and bicriteria problems. Different approaches have been proposed and different dual problems have been formulated accordingly.

In this chapter we refer to the so called Mond-Wier type duality, Wolfe type duality and Mixed type duality. The classical duality results stated by Wolf [39] have been extended to the vector case by Weir in [37] under both convexity and pseudoconvexity assumptions. In [29] Mond and Weir introduced a scalar dual problem under pseudoconvexity (pseudoconcavity) assumptions and his duality results have been extended to the vector case by Egudo in [17].

---

*Department of Statistics and Applied Mathematics, Faculty of Economics, University of Pisa, Via Cosimo Ridolfi 10, 56124 Pisa (Italy). E-mails: cambric@ec.unipi.it - lcarosi@ec.unipi.it

In fact, over the last few years, there has been a very strict relationship between both Mond-Weir type and Wolfe type duality from one side, and generalized convexity from the other side; many researchers have addressed their attention in finding new classes of generalized convex (concave) functions which are useful tools to prove duality theorems. In this line invexity and generalized invexity (see for all [1, 3, 6, 7, 18, 19, 23, 26, 33]) as well as generalized type I and generalized (F, ρ)-convex function (see for all [5, 17, 21, 22, 24, 32, 36]) have been introduced and widely exploited in order to get duality results.

As a unified framework for the study of both Mond-Wier and Wolfe type duality, mixed type duality has been introduced by Xu [40]. He considers a primal problem where the feasible region is defined only by inequality constraint while more recent papers [2, 3, 22, 28, 31] suggest mixed type dual problems for a primal vector optimization problem whose feasible region is defined by equality and inequality constraints. Of course, even in the mixed type duality context, duality results are obtained under suitable generalized convexity assumptions ([1]).

At the best of our knowledge, the most part of the recent literature on duality deals with primal problems whose feasible region are defined by equality and inequality constraints and the related sufficient optimality conditions are Karush-Kuhn-Tucker type conditions. Few contributions deal with problems having a compact constrained feasible region (for this latter case the reader can see for example the leading article by Tanino and Sawaragi [35]).

In this chapter we aim to consider a primal program where the feasible region is defined by equality constraints, inequality and a set constraint; the set constraint is not required, a priori, to be open, closed or convex. The presence of the set constraint imposes to leave Karush-Kuhn-Tucker necessary conditions and to use some other necessary optimality conditions which can be classified as maximum principle conditions (see for example [14, 20]). Besides that, we assume that the image space of the primal objective function is ordered by an arbitrary convex, closed and pointed cone C, which is not necessarily the Paretian cone (the Paretian cone is the standard assumption in the literature): therefore the definition of efficiency as well as the generalized concavity properties we are going to use for duality results are referred to a cone C (see for example [9, 11, 12, 13]).

In this context we will introduce two "general" dual problems that can be classified as mixed type ones. In the first dual problem, called $D_\mathcal{S}$, the objective function is scalar, while the second one is called $D_\mathcal{V}$ and it has a vector valued objective function. In both cases, specifying the value of the parameters in the dual problem, we can analyze both Wolfe-type and Mond-Weir-type duality. Duality results are proved under suitable generalized concavity assumptions. Outside the mixed type context, a comparison between Wolfe-type and Mond-Weir-type duals can be found in [10].

It is worth noticing that the case where the feasible region is defined by just equality and inequality constraints can be seen as a particular case of our dual programs, that is

---

[1]The concept of mixed type duality has been also extended to the class of multiobjective variational problem in [30].

the case where the set constraint is open. In order to specify our duality results, in this latter case we introduce new classes of generalized concave functions. The new properties are related to the well known generalized invexity, even if no functional $\eta(x, y)$ is used at all in the definitions. This makes these new introduced classes of functions easier to be used than the generalized invex ones. The relationship between the new properties and the invexity ones is remarked (see also [15]).

The chapter is organized as follows. In section 2 we state the problem, we recall all the standard definitions we are going to use throughout the chapter and we present the so called necessary maximum principle condition. Furthermore, we introduce a new class of generalized concave function to be used in obtaining optimality conditions and duality results. In section 3 we specify how some generalized concavity conditions transform the necessary maximum principle condition as a sufficient condition. In section 4 we first introduce the two mixed type dual ploblems and the weak and strong duality results are presented for both programs. The particular case where the feasible region is defined by just equality and inequality constraints is studied in section 5. Section 6 presents the conclusions of the chapter.

2   Definitions and preliminary results

The aim of this section is to introduce the primal problem which will be studied through the whole chapter, to provide the necessary optimality conditions which will represent the basis for defining the dual problems, and to introduce a new class of generalized concave functions which will be useful in stating duality results.

2.1   Primal problem

The aim of this chapter is to study optimality results for vector optimization problems having both inequality and equality constraints as well as a set constraint (which covers the constraints which cannot be expressed by means of neither equalities nor inequalities). In particular, we will consider the following class of problems:

$$P: \begin{cases} C^*\text{-}\max \quad f(x) \\ \quad g(x) \in V \qquad inequality\ constraints \\ \quad h(x) = 0 \qquad equality\ constraints \\ \quad x \in X \qquad\qquad set\ constraint \end{cases}$$

where $A \subseteq \Re^n$ is a convex open set, $f : A \to \Re^s$, $g : A \to \Re^m$ and $h : A \to \Re^p$ are Fréchet differentiable vector valued functions with $s \geq 1$ and $m, p \geq 0$, $C \subset \Re^s$ and $V \subset \Re^m$ are closed convex pointed cones with nonempty interior (that is to say convex pointed solid cones), C^* is a cone such that $C^* = C$ or $Int(C) \subseteq C^* \subseteq C \setminus \{0\}$. The set $X \subseteq A$ is not required to be convex, closed or open.

For the sake of convenience, P can be also expressed as

$$P : \begin{cases} C^*\text{-}\max \; f(x) \\ \quad x \in S_P \end{cases}$$

where $S_P = \{x \in A : g(x) \in V, h(x) = 0, x \in X\}$.

Given the cone C^*, a feasible point $x_0 \in S_P$ is said to be a *local C^*-efficient point* if there exists a suitable neighbourhood I_{x_0} of x_0 such that:

$$\nexists y \in I_{x_0} \cap S_P, \; y \neq x_0, \;\; \text{such that } f(y) \in f(x_0) + C^*$$

or, in other words, such that:

$$f(y) \notin f(x_0) + C^* \quad \forall y \in I_{x_0} \cap S_P, \; y \neq x_0$$

In particular, $x_0 \in S_P$ is said to be a *local efficient point* in the case $C^* = C \setminus \{0\}$, it is said to be a *local weak efficient point* in the case $C^* = Int(C)$, it is said to be a *local strict efficient point* in the case $C^* = C$. Obviously, the previous definitions will be *global* instead of just local in the case the corresponding property holds for $I_{x_0} = A$.

2.2 Necessary optimality conditions

The following necessary optimality conditions of the maximum principle type has been proved in [14] as a particular case of more general results. Maximum principle type necessary optimality conditions for vector valued problems have been studied also in [20, 25]. In this light, notice that C^+ and V^+ represent the positive polar cones of the cones C and V, respectively.

Theorem 1 *Consider problem P, assume functions f, g and h to be Fréchet differentiable at $x_0 \in S_P$ and let h be continuous on a neighborhood of x_0. If X is convex and the feasible point $x_0 \in S_P$ is a local efficient point then the following maximum principle condition holds:*

- $\exists (\alpha_f, \alpha_g, \alpha_h) \in (C^+ \times V^+ \times \Re^p)$, $(\alpha_f, \alpha_g, \alpha_h) \neq 0$, *such that:*

$$\alpha_g^T g(x_0) = 0 \quad and \quad [\alpha_f^T J_f(x_0) + \alpha_g^T J_g(x_0) + \alpha_h^T J_h(x_0)](y - x_0) \leq 0 \;\; \forall y \in X$$

Assuming suitable additional constraints qualification conditions, the previous maximum principle necessary optimality condition holds with $\alpha_f \neq 0$.

The previous maximum principle necessary optimality condition suggests the introduction of two kinds of lagrangean functions, a scalar and a vector ones, which will be used in the rest of the chapter in order to state sufficient optimality conditions and duality results by means of the use of suitable scalar or vector generalized concavity properties.

Definition 1 *Given problem P and given a vector $c \in Int(C)$, the following Lagrangean type functions can be defined:*

- $\mathcal{L}_{\mathcal{S}} : (A \times C^+ \times V^+ \times \Re^p) \to \Re$ *such that:*

$$\mathcal{L}_{\mathcal{S}}(x, \alpha_f, \alpha_g, \alpha_h) = \alpha_f^T f(x) + \alpha_g^T g(x) + \alpha_h^T h(x)$$

- $\mathcal{L}_{\mathcal{V}} : (A \times C^+ \setminus \{0\} \times V^+ \times \Re^p) \to \Re^s$ *such that:*

$$\mathcal{L}_{\mathcal{V}}(x, \alpha_f, \alpha_g, \alpha_h) = f(x) + \frac{c}{\alpha_f^T c} \left[\alpha_g^T g(x) + \alpha_h^T h(x) \right]$$

Notice that the gradient of the scalar lagrangean function $\mathcal{L}_{\mathcal{S}}$, with respect to variable x, is:

$$\nabla \mathcal{L}_{\mathcal{S}}(x, \alpha_f, \alpha_g, \alpha_h)^T = \alpha_f^T J_f(x) + \alpha_g^T J_g(x) + \alpha_h^T J_h(x)$$

while the Jacobian of the vector valued lagrangean function $\mathcal{L}_{\mathcal{V}}$, with respect to variable x, results to be:

$$J_{\mathcal{L}_{\mathcal{V}}}(x, \alpha_f, \alpha_g, \alpha_h) = J_f(x) + \frac{c}{\alpha_f^T c} \left[\alpha_g^T J_g(x) + \alpha_h^T J_h(x) \right]$$

Notice that if $\alpha_f \neq 0$ then $\alpha_f^T J_{\mathcal{L}_{\mathcal{V}}}(x, \alpha_f, \alpha_g, \alpha_h) = \nabla \mathcal{L}_{\mathcal{S}}(x, \alpha_f, \alpha_g, \alpha_h)^T$.

In this light, the maximum principle condition given in Theorem 1 can be rewritten as follows:

- $\exists (\alpha_f, \alpha_g, \alpha_h) \in (C^+ \times V^+ \times \Re^p)$, $(\alpha_f, \alpha_g, \alpha_h) \neq 0$, such that:

$$\alpha_g^T g(x_0) = 0 \quad \text{and} \quad \nabla \mathcal{L}_{\mathcal{S}}(x_0, \alpha_f, \alpha_g, \alpha_h)^T (y - x_0) \leq 0 \quad \forall y \in X$$

For the sake of completeness, just notice that in the case X is an open set then the previous maximum principle necessary optimality condition reduces to the classical Fritz John one:

- $\exists (\alpha_f, \alpha_g, \alpha_h) \in (C^+ \times V^+ \times \Re^p)$, $(\alpha_f, \alpha_g, \alpha_h) \neq 0$, such that:

$$\alpha_g^T g(x_0) = 0 \quad \text{and} \quad \nabla \mathcal{L}_{\mathcal{S}}(x_0, \alpha_f, \alpha_g, \alpha_h) = 0$$

that is nothing but the Karush-Kuhn-Tucker condition in the case a constraint qualification condition holds (and $\alpha_f \neq 0$ is guaranteed).

2.3   Generalized concavity properties

The sufficient optimality conditions and the duality results which are provided in the forthcoming sections are based on suitable generalized concavity assumptions. For this very reason, it is worth recalling some known generalized concavity definitions, concerning both scalar and vector functions (see for all [4, 8]).

Definition 2 *A differentiable scalar function $f : A \to \Re$, $A \subseteq \Re^n$ open convex set, is said to be quasiconcave if:*

$$f(x_1) \geq f(x_2) \quad \Rightarrow \quad \nabla f(x_2)^T(x_1 - x_2) \geq 0 \qquad \forall x_1, x_2 \in A,$$

it is said to be pseudoconcave if:

$$f(x_1) > f(x_2) \quad \Rightarrow \quad \nabla f(x_2)^T(x_1 - x_2) > 0 \qquad \forall x_1, x_2 \in A,$$

while it is said to be strictly pseudoconcave if:

$$f(x_1) \geq f(x_2) \quad \Rightarrow \quad \nabla f(x_2)^T(x_1 - x_2) > 0 \qquad \forall x_1, x_2 \in A, \ x_1 \neq x_2.$$

Definition 3 *Let $C \subset \Re^s$ be a closed convex pointed cone with nonempty interior, and let $C^* \subseteq C$ be a cone such that $C^* = C$ or $Int(C) \subseteq C^* \subseteq C \setminus \{0\}$. A differentiable vector valued function $f : A \to \Re^s$, $A \subseteq \Re^n$ open convex set, is said to be C^*-quasiconcave if:*

$$f(x_1) \in f(x_2) + C^* \quad \Rightarrow \quad J_f(x_2)(x_1 - x_2) \in C \qquad \forall x_1, x_2 \in A, \ x_1 \neq x_2,$$

it is said to be weakly C^*-pseudoconcave if:

$$f(x_1) \in f(x_2) + C^* \quad \Rightarrow \quad J_f(x_2)(x_1 - x_2) \in C \setminus \{0\} \qquad \forall x_1, x_2 \in A, \ x_1 \neq x_2.$$

while it is said to be C^*-pseudoconcave if:

$$f(x_1) \in f(x_2) + C^* \quad \Rightarrow \quad J_f(x_2)(x_1 - x_2) \in Int(C) \qquad \forall x_1, x_2 \in A, \ x_1 \neq x_2.$$

The generalized concavity properties recalled in Definitions 2 and 3 are said to be verified "at $x_0 \in A$" in the case the corresponding conditions hold for $x_2 = x_0$.

The previously recalled generalized concavity properties will be useful in studying sufficient optimality conditions and duality results related to the maximum principle necessary optimality conditions. In order to study results based on the Fritz John necessary optimality conditions the following new classes of generalized concave functions are introduced.

Definition 4 *A differentiable scalar function $f : A \to \Re$, $A \subseteq \Re^n$ open convex set, is said to be semi-pseudoconcave in A if for all $x_1, x_2 \in A$ it holds:*

$$f(x_1) > f(x_2) \quad \Rightarrow \quad \nabla f(x_2) \neq 0 \qquad (1)$$

while it is said to be strictly semi-pseudoconcave in A if for all $x_1, x_2 \in A$, $x_1 \neq x_2$, it holds:

$$f(x_1) \geq f(x_2) \quad \Rightarrow \quad \nabla f(x_2) \neq 0 \qquad (2)$$

As usual, the semi-pseudoconcavity properties introduced in Definition 4 are said to be verified *"at $x_0 \in A$"* in the case the corresponding conditions hold for $x_2 = x_0$.

Notice that the class of semi-pseudoconcave functions contains both the pseudoconcave functions and the η-pseudoconcave function [2]. Notice also that all the functions having no stationary points are semi-pseudoconcave. The following example points out that the class of semi-pseudoconcave functions does not coincide with the one of pseudoconcave functions.

Example 1 *Function $f(x_1, x_2) = x_1^2 + x_2$ is semi-pseudoconcave in $\Re^2$ since $\nabla f(x_1, x_2)^T = (2x_1, 1) \neq 0$ for all $(x_1, x_2) \in \Re^2$. On the other hand, function f is not pseudoconcave since $f(1,0) > f(0,0)$ while $\nabla f(0,0)^T((1,0) - (0,0)) = 0 \not> 0$.*

In [15], Caprari characterizes the ρ-pseudo invex functions and the pseudo (F, ρ)-convex functions by means of conditions analogous to (1) and (2). Notice that, unlike generalized invexity, the definition of semi-pseudoconcavity does not require any additional functional, thus resulting easier to be used and checked than the invex one. The following theorem points out the behavior of semi-pseudoconcave functions with respect to maximum points.

Theorem 2 *Let $f : A \to \Re$, $A \subseteq \Re^n$ open set, be a differentiable scalar function. The following properties are equivalent:*

i) *f is [strictly] semi-pseudoconcave in A;*

ii) *for all $x_0 \in A$ it holds: $\nabla f(x_0) = 0$ if and only if x_0 is a [the unique] global maximum point.*

Proof i)$\Rightarrow$ii) Let f be [strictly] semi-pseudoconcave in A and let $x_0 \in A$. If x_0 is a global maximum point then the openess of A implies that $\nabla f(x_0) = 0$. If $\nabla f(x_0) = 0$ then the semi-pseudoconcavity of f implies that $f(x) \leq f(x_0)$ $[f(x) < f(x_0)]$ for all $x \in A$ and hence x_0 is a [the unique] global maximum point.
ii)$\Rightarrow$i) Assume by contradiction that $\exists x_1, x_2 \in A$ such that $f(x_1) > f(x_2)$ $[f(x_1) \geq f(x_2)]$ and $\nabla f(x_2) = 0$. For *ii)* x_2 is a [the unique] global maximum point and this contradicts the inequality $f(x_1) > f(x_2)$ $[f(x_1) \geq f(x_2)]$. $\square$

3 Sufficient optimality conditions

The aim of this section is to show how the necessary optimality conditions recalled in Subsection 2.2 become sufficient assuming suitable generalized concavity properties.

The following fundamental lemma holds. Notice that this lemma uses no generalized concavity assumption.

---

[2]Let $f(y) > f(x)$. If f is pseudoconcave then $\nabla f(x)^T(y - x) > 0$ and hence $\nabla f(x) \neq 0$. If f is η-pseudoconcave then $\nabla f(x)^T \eta(x, y) > 0$ and again $\nabla f(x) \neq 0$.

Lemma 1 *Consider problem P and assume functions f, g and h to be Fréchet differentiable. Let also $x_0, y \in S_P$ and $(\alpha_f, \alpha_g, \alpha_h) \in (C^+ \times V^+ \times \Re^p)$, $(\alpha_f, \alpha_g, \alpha_h) \neq 0$, be such that $\alpha_g^T g(x_0) = 0$. Then, the following properties hold:*

i) if $f(y) \in f(x_0) + C$ then:

$$\mathcal{L}_\mathcal{S}(y, \alpha_f, \alpha_g, \alpha_h) \geq \mathcal{L}_\mathcal{S}(x_0, \alpha_f, \alpha_g, \alpha_h)$$

ii) if $\alpha_f \neq 0$ and $f(y) \in f(x_0) + Int(C)$ then:

$$\mathcal{L}_\mathcal{S}(y, \alpha_f, \alpha_g, \alpha_h) > \mathcal{L}_\mathcal{S}(x_0, \alpha_f, \alpha_g, \alpha_h)$$

iii) if $f(y) \in f(x_0) + C^*$, where C^* is a cone such that $C^* = C$ or $Int(C) \subseteq C^* \subseteq C \backslash \{0\}$, then:

$$\mathcal{L}_\mathcal{V}(y, \alpha_f, \alpha_g, \alpha_h) \in \mathcal{L}_\mathcal{V}(x_0, \alpha_f, \alpha_g, \alpha_h) + C^*$$

Proof First notice that for the hypotheses it is $h(x_0) = 0$, so that $\alpha_h^T h(x_0) = 0$, and $\alpha_g^T g(x_0) = 0$. As a consequence it results:

$$\mathcal{L}_\mathcal{S}(x_0, \alpha_f, \alpha_g, \alpha_h) = \alpha_f^T f(x_0) \quad \text{and} \quad \mathcal{L}_\mathcal{V}(x_0, \alpha_f, \alpha_g, \alpha_h) = f(x_0) \qquad (3)$$

Let us now prove property *i)*. Since $y \in S_P$, $(\alpha_f, \alpha_g, \alpha_h) \in (C^+ \times V^+ \times \Re^p)$ and $f(y) \in f(x_0) + C$ it yields:

$$\alpha_g^T g(y) \geq 0, \ \alpha_h^T h(y) = 0, \ \alpha_f^T f(y) \geq \alpha_f^T f(x_0)$$

The result then follows trivially by means of (3).

Property *ii)* can be proved analogously to *i)*. With this aim, just recall that since C is a closed convex pointed cone with nonempty interior then conditions $\alpha_f \neq 0$ and $f(y) \in f(x_0) + Int(C)$ imply that $\alpha_f^T f(y) > \alpha_f^T f(x_0)$.

Let us finally prove property *iii)*. Since $y \in S_P$ and $(\alpha_f, \alpha_g, \alpha_h) \in (C^+ \times V^+ \times \Re^p)$ it yields $\alpha_g^T g(y) \geq 0$ and $\alpha_h^T h(y) = 0$, hence $\mathcal{L}_\mathcal{V}(y, \alpha_f, \alpha_g, \alpha_h) \in f(y) + C$. Being $f(y) \in f(x_0) + C^*$, with $C^* \subset C$, and being C a closed convex pointed cone with nonempty interior it then follows that $\mathcal{L}_\mathcal{V}(y, \alpha_f, \alpha_g, \alpha_h) \in f(x_0) + C^*$, which implies the thesis by means of (3). $\qquad \square$

The following sufficient optimality results, concerning the maximum principle condition, follow directly from the previous lemma by assuming suitable generalized concavity properties.

Theorem 3 *Consider problem P and assume functions f, g and h to be Fréchet differentiable. Let also $x_0 \in S_P$ be a feasible point such that $\exists (\alpha_f, \alpha_g, \alpha_h) \in (C^+ \times V^+ \times \Re^p)$, $(\alpha_f, \alpha_g, \alpha_h) \neq 0$, with $\alpha_g^T g(x_0) = 0$ and*

$$\nabla \mathcal{L}_\mathcal{S}(x_0, \alpha_f, \alpha_g, \alpha_h)^T (y - x_0) \leq 0 \ \ \forall y \in X \qquad (4)$$

Then, the following properties hold:

i) *if function $\mathcal{L}_{\mathcal{S}}(x, \alpha_f, \alpha_g, \alpha_h)$ is strictly pseudoconcave at x_0 with respect to variable x then x_0 is a global strict efficient point;*

ii) *if $\alpha_f \neq 0$ and function $\mathcal{L}_{\mathcal{S}}(x, \alpha_f, \alpha_g, \alpha_h)$ is pseudoconcave at x_0 with respect to variable x then x_0 is a global weak efficient point;*

iii) *if $\alpha_f \neq 0$ and function $\mathcal{L}_{\mathcal{V}}(x, \alpha_f, \alpha_g, \alpha_h)$ is C^*-pseudoconcave at x_0 with respect to variable x then x_0 is a global C^*-efficient point, where C^* is a cone such that $C^* = C$ or $Int(C) \subseteq C^* \subseteq C \setminus \{0\}$.*

Proof Let us first prove property *i)* and assume by contradiction that there exists $y \in S_P$, $y \neq x_0$, such that $f(y) \in f(x_0) + C$. For *i)* of Lemma 1 it is:

$$\mathcal{L}_{\mathcal{S}}(y, \alpha_f, \alpha_g, \alpha_h) \geq \mathcal{L}_{\mathcal{S}}(x_0, \alpha_f, \alpha_g, \alpha_h)$$

The strict pseudoconcavity of $\mathcal{L}_{\mathcal{S}}(x, \alpha_f, \alpha_g, \alpha_h)$ at x_0 with respect to x then yields:

$$\nabla \mathcal{L}_{\mathcal{S}}(x_0, \alpha_f, \alpha_g, \alpha_h)^T (y - x_0) > 0$$

which contradicts (4).

Property *ii)* can be proved analogously to *i)*.

Let us finally prove property *iii)* and assume by contradiction that there exists $y \in S_P$, $y \neq x_0$, such that $f(y) \in f(x_0) + C^*$. For *iii)* of Lemma 1 it is:

$$\mathcal{L}_{\mathcal{V}}(y, \alpha_f, \alpha_g, \alpha_h) \in \mathcal{L}_{\mathcal{V}}(x_0, \alpha_f, \alpha_g, \alpha_h) + C^*$$

Since $\mathcal{L}_{\mathcal{V}}(x, \alpha_f, \alpha_g, \alpha_h)$ is $(C^*, Int(C))$-pseudoconcave at x_0 with respect to variable x it yields $J_{\mathcal{L}_{\mathcal{V}}}(x_0, \alpha_f, \alpha_g, \alpha_h)(y - x_0) \in Int(C)$, so that $\alpha_f \in C^+$, $\alpha_f \neq 0$, implies:

$$\nabla \mathcal{L}_{\mathcal{S}}(x_0, \alpha_f, \alpha_g, \alpha_h)^T (y - x_0) = \alpha_f^T J_{\mathcal{L}_{\mathcal{V}}}(x_0, \alpha_f, \alpha_g, \alpha_h)(y - x_0) > 0$$

which contradicts (4). □

Theorem 3 points out the key role played by the pseudoconcavity of $\mathcal{L}_{\mathcal{S}}(x, \alpha_f, \alpha_g, \alpha_h)$ and by the C^*-pseudoconcavity property of $\mathcal{L}_{\mathcal{V}}(x, \alpha_f, \alpha_g, \alpha_h)$ in order to let the maximum principle conditions become sufficient optimality conditions.

Notice that Theorem 3 generalizes the various results appeared in the literature concerning vector valued problems having no set constraints. Notice also that in [20] results concerning vector problems with set constraints have been proved in the particular case of Paretian cone orderings.

This section is concluded providing sufficient optimality results concerning the Fritz John condition. These results are based on the newly introduced classes of semi-pseudoconcave functions.

Theorem 4 *Consider problem P and assume functions f, g and h to be Fréchet differentiable. Let also $x_0 \in S_P$ be a feasible point such that $\exists (\alpha_f, \alpha_g, \alpha_h) \in (C^+ \times V^+ \times \Re^p)$, $(\alpha_f, \alpha_g, \alpha_h) \neq 0$, with $\alpha_g^T g(x_0) = 0$ and*

$$\nabla \mathcal{L}_{\mathcal{S}}(x_0, \alpha_f, \alpha_g, \alpha_h) = 0 \tag{5}$$

Then, the following properties hold:

   i) *if function $\mathcal{L}_{\mathcal{S}}(x, \alpha_f, \alpha_g, \alpha_h)$ is strictly semi-pseudoconcave at x_0 with respect to variable x then x_0 is a global strict efficient point;*

  ii) *if $\alpha_f \neq 0$ and function $\mathcal{L}_{\mathcal{S}}(x, \alpha_f, \alpha_g, \alpha_h)$ is semi-pseudoconcave at x_0 with respect to variable x then x_0 is a global weak efficient point.*

Proof Let us first prove property *i)* and assume by contradiction that there exists $y \in S_P$, $y \neq x_0$, such that $f(y) \in f(x_0) + C$. For *i)* of Lemma 1 it is:

$$\mathcal{L}_{\mathcal{S}}(y, \alpha_f, \alpha_g, \alpha_h) \geq \mathcal{L}_{\mathcal{S}}(x_0, \alpha_f, \alpha_g, \alpha_h)$$

The strict semi-pseudoconcavity of $\mathcal{L}_{\mathcal{S}}(x, \alpha_f, \alpha_g, \alpha_h)$ at x_0 with respect to x then yields:

$$\nabla \mathcal{L}_{\mathcal{S}}(x_0, \alpha_f, \alpha_g, \alpha_h) \neq 0$$

which contradicts (5).

    Property *ii)* can be proved analogously to *i)*. $\square$

4   Duality results for set constrained problems

The aim of this section is to study duality results for problem P, which from now on will be referred to as the primal problem.

4.1   Scalar and vector dual problems

By following two classical approaches of the literature, the following dual problems can be defined by using the previously stated maximum principle conditions (notice that in the Wolfe type dual it is $c \in Int(C)$):

Mond-Weir type Dual

$$\begin{cases} C\text{-}\min \quad f(x) \\ \alpha_g^T g(x) \leq 0 \ , \ \alpha_h^T h(x) = 0 \\ \nabla \mathcal{L}_{\mathcal{S}}(x, \alpha_f, \alpha_g, \alpha_h)^T (y - x) \leq 0 \ \forall y \in X, \\ x \in A, \ \alpha_f \in C^+ \setminus \{0\}, \ \alpha_g \in V^+, \ \alpha_h \in \Re^p \end{cases}$$

Wolfe type Dual

$$
\begin{cases}
C_-\min \ f(x) + \frac{c}{\alpha_f^T c}[\alpha_g^T g(x) + \alpha_h^T h(x)] \\
\nabla \mathcal{L}_\mathcal{S}(x, \alpha_f, \alpha_g, \alpha_h)^T (y - x) \leq 0 \ \forall y \in X, \\
x \in A, \ \alpha_f \in C^+ \setminus \{0\}, \ \alpha_g \in V^+, \ \alpha_h \in \Re^p
\end{cases}
$$

A comparison of these two dual problems points out that the easier is the objective function, the more complex are the constraints, and *vice versa*.

Another approach, named mixed duality, has been proposed in the literature in order to manage in an unifying framework both the Mond-Weir and the Wolfe duals. With this aim the following notations are needed:

- $\delta \in \{0, 1\}$ is a $0 - 1$ parameter;

- $\mathcal{J} = \{J_1, J_2, J_3, J_4\}$ is a partition of $\mathcal{P} = \{1, \ldots, p\}$;

- $h(x) = [h_1(x), h_2(x), h_3(x), h_4(x)]$ and $\alpha_h = (\alpha_{h_1}, \alpha_{h_2}, \alpha_{h_3}, \alpha_{h_4})$ are partitioned accordingly to $\mathcal{J}$;

- $\Delta(x, \alpha_g, \alpha_{h_2}, \alpha_{h_3}, \alpha_{h_4}) = \delta \alpha_g^T g(x) + \alpha_{h_2}^T h_2(x) + \alpha_{h_3}^T h_3(x) + \alpha_{h_4}^T h_4(x)$;

The two following functions can then be defined:

$$
\begin{aligned}
\mathcal{F}_\mathcal{S}(x, \alpha_f, \alpha_g, \alpha_h) &= \mathcal{L}_\mathcal{S}(x, \alpha_f, \alpha_g, \alpha_h) - \Delta(x, \alpha_g, \alpha_{h_2}, \alpha_{h_3}, \alpha_{h_4}) \\
&= \alpha_f^T f(x) + (1 - \delta)\alpha_g^T g(x) + \alpha_{h_1}^T h_1(x) \\
\mathcal{F}_\mathcal{V}(x, \alpha_f, \alpha_g, \alpha_h) &= \mathcal{L}_\mathcal{V}(x, \alpha_f, \alpha_g, \alpha_h) - \frac{c}{\alpha_f^T c}\Delta(x, \alpha_g, \alpha_{h_2}, \alpha_{h_3}, \alpha_{h_4}) \\
&= f(x) + \frac{c}{\alpha_f^T c}\left[(1 - \delta)\alpha_g^T g(x) + \alpha_{h_1}^T h_1(x)\right]
\end{aligned}
$$

Notice that $\mathcal{F}_\mathcal{S}(x, \alpha_f, \alpha_g, \alpha_h)$ is a scalar function while $\mathcal{F}_\mathcal{V}(x, \alpha_f, \alpha_g, \alpha_h)$ is a vector one. Notice also that if $\alpha_f \neq 0$ then $\mathcal{F}_\mathcal{S}(x, \alpha_f, \alpha_g, \alpha_h) = \alpha_f^T \mathcal{F}_\mathcal{V}(x, \alpha_f, \alpha_g, \alpha_h)$.

The following mixed type dual problems $D_\mathcal{S}$ and $D_\mathcal{V}$ will be referred to as the scalar and the vector mixed duals of P, respectively ([3]):

Mixed type Duals

$$
D_\mathcal{S} : \begin{cases}
\min \ \mathcal{F}_\mathcal{S}(x, \alpha_f, \alpha_g, \alpha_h) \\
\nabla \mathcal{L}_\mathcal{S}(x, \alpha_f, \alpha_g, \alpha_h)^T (y - x) \leq 0 \ \forall y \in X, \\
(x, \alpha_f, \alpha_g, \alpha_h) \in S_D
\end{cases}
$$

---

[3]Notice that mixed type dual problems have been first introduced by Xu in [40] in the case of paretian cones C and V and a primal feasible region defined by just inequality constraints. Similar results for a primal feasible region defined by both equality and inequality constraints have been proposed in [2, 3, 22, 28, 31].

$$D_{\mathcal{V}} : \begin{cases} C^*\_\min \;\; \mathcal{F}_{\mathcal{V}}(x, \alpha_f, \alpha_g, \alpha_h) \\ \nabla \mathcal{L}_{\mathcal{S}}(x, \alpha_f, \alpha_g, \alpha_h)^T (y - x) \leq 0 \; \forall y \in X, \\ \alpha_f \neq 0 \\ (x, \alpha_f, \alpha_g, \alpha_h) \in S_D \end{cases}$$

where:

$$S_D : \begin{cases} (x, \alpha_f, \alpha_g, \alpha_h) \;\; \text{such that} \\ \delta \alpha_g^T g(x) + \alpha_{h_2}^T h_2(x) \leq 0 \\ \alpha_{h_3}^T h_3(x) = 0, \;\; \alpha_{h_4}^T h_4(x) \leq 0 \\ x \in A, \;\; \alpha_f \in C^+, \;\; \alpha_g \in V^+, \;\; \alpha_h \in \Re^p \end{cases}$$

Notice that both the two previous dual problems are based on the maximum principle condition:

$$\nabla \mathcal{L}_{\mathcal{S}}(x_0, \alpha_f, \alpha_g, \alpha_h)^T (y - x_0) \leq 0 \;\; \forall y \in X \qquad (6)$$

It is worth noticing also that the Mond-Weir type dual previously described is nothing but the particular case of $D_{\mathcal{V}}$ where $\delta = 1$, $J_3 = \{1, \ldots, p\}$ and $J_1 = J_2 = J_4 = \emptyset$. On the other hand, the previous Wolfe type dual can be obtained from $D_{\mathcal{V}}$ assuming $\delta = 0$, $J_1 = \{1, \ldots, p\}$ and $J_2 = J_3 = J_4 = \emptyset$.

Obviously, a point $(x_0, \alpha_f, \alpha_g, \alpha_h)$ feasible for $D_{\mathcal{V}}$ is said to be a *global C^*-minimum point* for $D_{\mathcal{V}}$ if:

$$\nexists (\widehat{x}, \widehat{\alpha}_f, \widehat{\alpha}_g, \widehat{\alpha}_h) \;\; \text{feasible for } D_{\mathcal{V}} \text{ such that } f(\widehat{x}, \widehat{\alpha}_f, \widehat{\alpha}_g, \widehat{\alpha}_h) \in f(x_0, \alpha_f, \alpha_g, \alpha_h) - C^*$$

or, in other words, if:

$$f(x_0, \alpha_f, \alpha_g, \alpha_h) \notin f(\widehat{x}, \widehat{\alpha}_f, \widehat{\alpha}_g, \widehat{\alpha}_h) + C^* \;\; \forall (\widehat{x}, \widehat{\alpha}_f, \widehat{\alpha}_g, \widehat{\alpha}_h) \;\; \text{feasible for } \; D_{\mathcal{V}}$$

where, as usual, it is assumed $(\widehat{x}, \widehat{\alpha}_f, \widehat{\alpha}_g, \widehat{\alpha}_h) \neq (x_0, \alpha_f, \alpha_g, \alpha_h)$.

Duality results can be proved for P with respect to both the scalar dual $D_{\mathcal{S}}$ and the vector dual $D_{\mathcal{V}}$. As usual, these results will be proved assuming some suitable generalized concavity properties. Subsections 4.2 and 4.3 present weak and strong duality results for the scalar dual problem and the vector dual problem, respectively. Each weak duality theorem is preceded by a fundamental preliminary lemma where no generalized concavity assumption is involved.

4.2 Duality results for $D_{\mathcal{S}}$

Lemma 2 *Let us consider the primal problem P and the region S_D. Then, $\forall x_1 \in S_P$ and $\forall (x_2, \alpha_f, \alpha_g, \alpha_h) \in S_D$ it results:*

i) if $\alpha_f^T f(x_1) \geq \mathcal{F}_{\mathcal{S}}(x_2, \alpha_f, \alpha_g, \alpha_h)$ then $\mathcal{L}_{\mathcal{S}}(x_1, \alpha_f, \alpha_g, \alpha_h) \geq \mathcal{L}_{\mathcal{S}}(x_2, \alpha_f, \alpha_g, \alpha_h)$;

ii) if $\alpha_f^T f(x_1) > \mathcal{F}_{\mathcal{S}}(x_2, \alpha_f, \alpha_g, \alpha_h)$ then $\mathcal{L}_{\mathcal{S}}(x_1, \alpha_f, \alpha_g, \alpha_h) > \mathcal{L}_{\mathcal{S}}(x_2, \alpha_f, \alpha_g, \alpha_h)$.

Proof First notice that $\forall x_1 \in S_P$ and $\forall (x_2, \alpha_f, \alpha_g, \alpha_h) \in S_D$ it results:

$$\mathcal{L}_{\mathcal{S}}(x_1, \alpha_f, \alpha_g, \alpha_h) \geq \alpha_f^T f(x_1) \tag{7}$$

Since $(x_2, \alpha_f, \alpha_g, \alpha_h) \in S_D$ it is also $\Delta(x_2, \alpha_g, \alpha_{h_2}, \alpha_{h_3}, \alpha_{h_4}) \leq 0$ so that:

$$\mathcal{F}_{\mathcal{S}}(x_2, \alpha_f, \alpha_g, \alpha_h) \geq \mathcal{L}_{\mathcal{S}}(x_2, \alpha_f, \alpha_g, \alpha_h) \tag{8}$$

Both the two results then follows trivially from (7) and (8). $\qquad\square$

The weak duality result for the pair of problems P and $D_{\mathcal{S}}$ follows with the use of suitable pseudoconcavity assumptions.

Theorem 5 (Weak duality for $D_{\mathcal{S}}$) *Let us consider the primal problem P and the dual problem $D_{\mathcal{S}}$. Assume also that function $\mathcal{L}_{\mathcal{S}}(x, \alpha_f, \alpha_g, \alpha_h)$ is pseudoconcave with respect to variable x for all multipliers $(\alpha_f, \alpha_g, \alpha_h) \in (C^+ \times V^+ \times \Re^p)$. Then, $\forall x_1 \in S_P$ and $\forall (x_2, \alpha_f, \alpha_g, \alpha_h) \in S_D$ verifying (6) it is:*

$$\alpha_f^T f(x_1) \leq \mathcal{F}_{\mathcal{S}}(x_2, \alpha_f, \alpha_g, \alpha_h)$$

Proof Suppose by contradiction that $\exists x_1 \in S_P$ and $\exists (x_2, \alpha_f, \alpha_g, \alpha_h) \in S_D$ such that $\alpha_f^T f(x_1) > \mathcal{F}_{\mathcal{S}}(x_2, \alpha_f, \alpha_g, \alpha_h)$. Then, *ii)* of Lemma 2 yields:

$$\mathcal{L}_{\mathcal{S}}(x_1, \alpha_f, \alpha_g, \alpha_h) > \mathcal{L}_{\mathcal{S}}(x_2, \alpha_f, \alpha_g, \alpha_h)$$

and hence, from the pseudoconcavity of $\mathcal{L}_{\mathcal{S}}(x, \alpha_f, \alpha_g, \alpha_h)$, it results:

$$\nabla \mathcal{L}_{\mathcal{S}}(x_2, \alpha_f, \alpha_g, \alpha_h)^T (x_1 - x_2) > 0$$

which contradicts (6). $\qquad\square$

Corollary 1 *Let us consider the primal problem P and the dual problem $D_{\mathcal{S}}$. Assume also that function $\mathcal{L}_{\mathcal{S}}(x, \alpha_f, \alpha_g, \alpha_h)$ is pseudoconcave with respect to variable x for all multipliers $(\alpha_f, \alpha_g, \alpha_h) \in (C^+ \times V^+ \times \Re^p)$. If $x_0 \in S_P$ and $(x_0, \alpha_f, \alpha_g, \alpha_h)$ is feasible for $D_{\mathcal{S}}$, with $(1 - \delta)\alpha_g^T g(x_0) = 0$, then the following properties hold:*

i) $\alpha_f^T f(x_0) = \mathcal{F}_{\mathcal{S}}(x_0, \alpha_f, \alpha_g, \alpha_h)$;

ii) $(x_0, \alpha_f, \alpha_g, \alpha_h)$ is a global minimum for $D_{\mathcal{S}}$;

iii) if $\alpha_f \neq 0$ then x_0 is a global weak efficient point for P.

Proof *i)* Just notice that $x_0 \in S_P$ implies $h(x_0) = 0$ so that it results:

$$\mathcal{F}_{\mathcal{S}}(x_0, \alpha_f, \alpha_g, \alpha_h) - \alpha_f^T f(x_0) = (1 - \delta)\alpha_g^T g(x_0) = 0$$

ii) Suppose by contradiction that there exists $(\widehat{x}, \widehat{\alpha}_f, \widehat{\alpha}_g, \widehat{\alpha}_h)$ feasible for $D_{\mathcal{S}}$ such that $\mathcal{F}_{\mathcal{S}}(x_0, \alpha_f, \alpha_g, \alpha_h) > \mathcal{F}_{\mathcal{S}}(\widehat{x}, \widehat{\alpha}_f, \widehat{\alpha}_g, \widehat{\alpha}_h)$. From *i)* it then follows that $\alpha_f^T f(x_0) > \mathcal{F}_{\mathcal{S}}(\widehat{x}, \widehat{\alpha}_f, \widehat{\alpha}_g, \widehat{\alpha}_h)$ which contradicts the weak duality result.

iii) Suppose by contradiction that there exists $y \in S_P$ such that $f(y) \in f(x_0) + Int(C)$, so that $\alpha_f \neq 0$ yields $\alpha_f^T f(y) > \alpha_f^T f(x_0)$. From *i)* it then follows that $\alpha_f^T f(y) > \mathcal{F}_{\mathcal{S}}(x_0, \alpha_f, \alpha_g, \alpha_h)$ and this contradicts the weak duality result. □

By means of the maximum principle necessary optimality condition given in Theorem 1, it is now possible to prove the following strong duality results. Notice that in the following results the set X is required to be convex.

Theorem 6 (Strong duality for $D_{\mathcal{S}}$) *Let us consider the primal problem P and the dual problem $D_{\mathcal{S}}$. Assume also that function $\mathcal{L}_{\mathcal{S}}(x, \alpha_f, \alpha_g, \alpha_h)$ is pseudoconcave with respect to variable x for all multipliers $(\alpha_f, \alpha_g, \alpha_h) \in (C^+ \times V^+ \times \Re^p)$ and that X is convex. Then, for all $x_0 \in S_P$ global efficient point for P $\exists \alpha_f \in C^+$, $\exists \alpha_g \in V^+$, $\exists \alpha_h \in \Re^p$, $(\alpha_f, \alpha_g, \alpha_h) \neq 0$, such that:*

i) $\alpha_f^T f(x_0) = \mathcal{F}_{\mathcal{S}}(x_0, \alpha_f, \alpha_g, \alpha_h)$;

ii) $(x_0, \alpha_f, \alpha_g, \alpha_h)$ *is a global minimum point for $D_{\mathcal{S}}$.*

Proof Since x_0 is a global efficient point for P, then for Theorem 1 $\exists \alpha_f \in C^+$, $\exists \alpha_g \in V^+$, $\exists \alpha_h \in \Re^p$, $(\alpha_f, \alpha_g, \alpha_h) \neq 0$, such that $\alpha_g^T g(x_0) = 0$ and

$$\nabla \mathcal{L}_{\mathcal{S}}(x_0, \alpha_f, \alpha_g, \alpha_h)^T (y - x_0) \leq 0 \quad \forall y \in X$$

This implies that $x_0 \in S_P$, $(x_0, \alpha_f, \alpha_g, \alpha_h)$ is feasible for $D_{\mathcal{S}}$, and $(1 - \delta)\alpha_g^T g(x_0) = 0$. The results then follows from Corollary 1. □

The following further duality result follows from the strong duality theorem.

Corollary 2 *Let us consider the primal problem P and the dual problem $D_{\mathcal{S}}$. Assume also that function $\mathcal{L}_{\mathcal{S}}(x, \alpha_f, \alpha_g, \alpha_h)$ is pseudoconcave with respect to variable x for all multipliers $(\alpha_f, \alpha_g, \alpha_h) \in (C^+ \times V^+ \times \Re^p)$ and that X is convex. Then, for all $x_1 \in S_P$ global efficient point for P and for all $(x_2, \alpha_f, \alpha_g, \alpha_h)$ global minimum point for $D_{\mathcal{S}}$ it results:*

$$\exists \widehat{\alpha}_f \in C^+ \text{ such that } \widehat{\alpha}_f^T f(x_1) = \mathcal{F}_{\mathcal{S}}(x_2, \alpha_f, \alpha_g, \alpha_h)$$

Proof For the previous strong duality theorem $\exists \widehat{\alpha}_f \in C^+$, $\exists \widehat{\alpha}_g \in V^+$, $\exists \widehat{\alpha}_h \in \Re^p$, $(\widehat{\alpha}_f, \widehat{\alpha}_g, \widehat{\alpha}_h) \neq 0$, such that $\widehat{\alpha}_f^T f(x_1) = \mathcal{F}_{\mathcal{S}}(x_1, \widehat{\alpha}_f, \widehat{\alpha}_g, \widehat{\alpha}_h)$ and $(x_1, \widehat{\alpha}_f, \widehat{\alpha}_g, \widehat{\alpha}_h)$ is a global minimum point for $D_{\mathcal{S}}$. Since $(x_2, \alpha_f, \alpha_g, \alpha_h)$ is a global minimum point for $D_{\mathcal{S}}$ too, it results:

$$\mathcal{F}_{\mathcal{S}}(x_1, \widehat{\alpha}_f, \widehat{\alpha}_g, \widehat{\alpha}_h) = \mathcal{F}_{\mathcal{S}}(x_2, \alpha_f, \alpha_g, \alpha_h)$$

and the thesis is proved being $\widehat{\alpha}_f^T f(x_1) = \mathcal{F}_{\mathcal{S}}(x_1, \widehat{\alpha}_f, \widehat{\alpha}_g, \widehat{\alpha}_h)$.   □

4.3   Duality results for $\mathbf{D}_{\mathcal{V}}$

Lemma 3 *Let us consider the primal problem P and the region S_D, and let C^* be a cone such that $C^* = C$ or $Int(C) \subseteq C^* \subseteq C \setminus \{0\}$. Then, $\forall x_1 \in S_P$ and $\forall (x_2, \alpha_f, \alpha_g, \alpha_h) \in S_D$, $\alpha_f \neq 0$, such that $f(x_1) \in \mathcal{F}_{\mathcal{V}}(x_2, \alpha_f, \alpha_g, \alpha_h) + C^*$ it results:*

i) $\mathcal{L}_{\mathcal{V}}(x_1, \alpha_f, \alpha_g, \alpha_h) \in \mathcal{L}_{\mathcal{V}}(x_2, \alpha_f, \alpha_g, \alpha_h) + C^*$;

ii) $\mathcal{L}_{\mathcal{S}}(x_1, \alpha_f, \alpha_g, \alpha_h) \geq \mathcal{L}_{\mathcal{S}}(x_2, \alpha_f, \alpha_g, \alpha_h)$;

iii) if $C^* = Int(C)$ then $\mathcal{L}_{\mathcal{S}}(x_1, \alpha_f, \alpha_g, \alpha_h) > \mathcal{L}_{\mathcal{S}}(x_2, \alpha_f, \alpha_g, \alpha_h)$.

Proof *i)* First notice that for all $(\alpha_f, \alpha_g, \alpha_h) \in (C^+ \times V^+ \times \Re^p)$, $\alpha_f \neq 0$, and for all $x_1 \in S_P$ it results:

$$\mathcal{L}_{\mathcal{V}}(x_1, \alpha_f, \alpha_g, \alpha_h) \in f(x_1) + C$$

As a consequence, being $f(x_1) \in \mathcal{F}_{\mathcal{V}}(x_2, \alpha_f, \alpha_g, \alpha_h) + C^*$, the pointedness of the closed convex cone C implies that:

$$\mathcal{L}_{\mathcal{V}}(x_1, \alpha_f, \alpha_g, \alpha_h) \in \mathcal{F}_{\mathcal{V}}(x_2, \alpha_f, \alpha_g, \alpha_h) + C^* \tag{9}$$

By definition it is $\mathcal{F}_{\mathcal{V}}(x_2, \alpha_f, \alpha_g, \alpha_h) = \mathcal{L}_{\mathcal{V}}(x_2, \alpha_f, \alpha_g, \alpha_h) - \frac{c}{\alpha_f^T c} \Delta(x_2, \alpha_g, \alpha_{h_2}, \alpha_{h_3}, \alpha_{h_4})$. Since $\Delta(x_2, \alpha_g, \alpha_{h_2}, \alpha_{h_3}, \alpha_{h_4}) \leq 0$ holds it yields

$$\mathcal{F}_{\mathcal{V}}(x_2, \alpha_f, \alpha_g, \alpha_h) \in \mathcal{L}_{\mathcal{V}}(x_2, \alpha_f, \alpha_g, \alpha_h) + C \tag{10}$$

The result then follows from (9) and (10) being C a closed convex pointed solid cone.

  ii),iii) Follows from *i)* taking into account that $\alpha_f \neq 0$ and that $\mathcal{L}_{\mathcal{S}}(x, \alpha_f, \alpha_g, \alpha_h) = \alpha_f^T \mathcal{L}_{\mathcal{V}}(x, \alpha_f, \alpha_g, \alpha_h)$.   □

In the case of the pair of problems P and $D_{\mathcal{V}}$, duality results are obtained by means of suitable vector and scalar pseudoconcavity properties.

Theorem 7 (Weak duality for $\mathbf{D}_{\mathcal{V}}$) *Let us consider the primal problem P and the dual problem $D_{\mathcal{V}}$, and let C^* be a cone such that $Int(C) \subseteq C^* \subseteq C \setminus \{0\}$. Assume also that at least one of the following conditions (C_1), (C_2) and (C_3) is verified for all multipliers $(\alpha_f, \alpha_g, \alpha_h) \in (C^+ \times V^+ \times \Re^p)$, $\alpha_f \neq 0$:*

(C_1) *function* $\mathcal{L}_{\mathcal{V}}(x, \alpha_f, \alpha_g, \alpha_h)$ *is* C^*-*pseudoconcave with respect to variable* x;

(C_2) *function* $\mathcal{L}_{\mathcal{S}}(x, \alpha_f, \alpha_g, \alpha_h)$ *is strictly pseudoconcave with respect to variable* x;

(C_3) $C^* = Int(C)$ *and* $\mathcal{L}_{\mathcal{S}}(x, \alpha_f, \alpha_g, \alpha_h)$ *is pseudoconcave with respect to variable* x.

Then, $\forall x_1 \in S_P$ *and* $\forall (x_2, \alpha_f, \alpha_g, \alpha_h) \in S_D$ *verifying (6),* $\alpha_f \neq 0$, *it is:*

$$f(x_1) \notin \mathcal{F}_{\mathcal{V}}(x_2, \alpha_f, \alpha_g, \alpha_h) + C^*$$

Proof Suppose by contradiction that $\exists x_1 \in S_P$, $\exists (x_2, \alpha_f, \alpha_g, \alpha_h) \in S_D$, $\alpha_f \neq 0$, such that

$$f(x_1) \in \mathcal{F}_{\mathcal{V}}(x_2, \alpha_f, \alpha_g, \alpha_h) + C^*$$

Assume (C_1) holds, then for *i)* of Lemma 3 it is $\mathcal{L}_{\mathcal{V}}(x_1, \alpha_f, \alpha_g, \alpha_h) \in \mathcal{L}_{\mathcal{V}}(x_2, \alpha_f, \alpha_g, \alpha_h) + C^*$ and hence $J_{\mathcal{L}_{\mathcal{V}}}(x_2, \alpha_f, \alpha_g, \alpha_h)(x_1 - x_2) \in Int(C)$, so that $\alpha_f \in C^+$, $\alpha_f \neq 0$, implies:

$$\nabla \mathcal{L}_{\mathcal{S}}(x_2, \alpha_f, \alpha_g, \alpha_h)^T (x_1 - x_2) = \alpha_f^T J_{\mathcal{L}_{\mathcal{V}}}(x_2, \alpha_f, \alpha_g, \alpha_h)(x_1 - x_2) > 0$$

which contradicts (6).

The result under assumptions (C_2) and (C_3) follows analogously by means of *ii)* and *iii)* of Lemma 3, respectively. $\qquad\qquad\qquad\qquad\qquad\qquad\qquad\qquad\qquad\qquad\qquad\square$

Corollary 3 *Let us consider the primal problem* P *and the dual problem* $D_{\mathcal{V}}$, *and let* C^* *be a cone such that* $Int(C) \subseteq C^* \subseteq C \setminus \{0\}$. *Assume also that at least one of conditions* (C_1), (C_2) *and* (C_3) *is verified for all multipliers* $(\alpha_f, \alpha_g, \alpha_h) \in (C^+ \times V^+ \times \Re^p)$, $\alpha_f \neq 0$. *If* $x_0 \in S_P$ *and* $(x_0, \alpha_f, \alpha_g, \alpha_h)$ *is feasible for* $D_{\mathcal{V}}$, *with* $(1 - \delta)\alpha_g^T g(x_0) = 0$, *then the following properties hold:*

i) $f(x_0) = \mathcal{F}_{\mathcal{V}}(x_0, \alpha_f, \alpha_g, \alpha_h)$;

ii) $(x_0, \alpha_f, \alpha_g, \alpha_h)$ *is a global* C^*-*minimum point for* $D_{\mathcal{V}}$;

iii) x_0 *is a global* C^*-*efficient point for* P.

Proof i) Just notice that $x_0 \in S_P$ implies $h(x_0) = 0$ so that it results:

$$\mathcal{F}_{\mathcal{V}}(x_0, \alpha_f, \alpha_g, \alpha_h) - f(x_0) = \frac{c}{\alpha_f^T c}(1 - \delta)\alpha_g^T g(x_0) = 0$$

ii) Suppose by contradiction that there exists $(\widehat{x}, \widehat{\alpha}_f, \widehat{\alpha}_g, \widehat{\alpha}_h)$ feasible for $D_{\mathcal{V}}$ such that $\mathcal{F}_{\mathcal{V}}(x_0, \alpha_f, \alpha_g, \alpha_h) \in \mathcal{F}_{\mathcal{V}}(\widehat{x}, \widehat{\alpha}_f, \widehat{\alpha}_g, \widehat{\alpha}_h) + C^*$. From *i)* it then follows that $f(x_0) \in \mathcal{F}_{\mathcal{V}}(\widehat{x}, \widehat{\alpha}_f, \widehat{\alpha}_g, \widehat{\alpha}_h) + C^*$ which contradicts the weak duality result.

iii) Suppose by contradiction that there exists $y \in S_P$ such that $f(y) \in f(x_0) + C^*$. From *i)* it then follows that $f(y) \in \mathcal{F}_{\mathcal{V}}(x_0, \alpha_f, \alpha_g, \alpha_h) + C^*$ and this contradicts the weak duality result.   □

As we have already done for problem $D_{\mathcal{S}}$ we prove strong duality result by means of the maximum principle necessary optimality condition given in Theorem 1. In the next results the set X is required to be convex and a constraints qualification condition is needed in order to guarantee $\alpha_f \neq 0$.

Theorem 8 (Strong duality for $D_{\mathcal{V}}$) *Let us consider the primal problem P and the dual problem $D_{\mathcal{V}}$. Assume also that at least one of conditions (C_1) (with $C^* = C \setminus \{0\}$) and (C_2) is verified for all multipliers $(\alpha_f, \alpha_g, \alpha_h) \in (C^+ \times V^+ \times \Re^p)$, $\alpha_f \neq 0$, that X is convex and that a constraint qualification condition holds for problem P. Then, for all $x_0 \in S_P$ global efficient point for P $\exists \alpha_f \in C^+ \setminus \{0\}$, $\exists \alpha_g \in V^+$, $\exists \alpha_h \in \Re^p$, such that:*

i) $f(x_0) = \mathcal{F}_{\mathcal{V}}(x_0, \alpha_f, \alpha_g, \alpha_h)$;

ii) $(x_0, \alpha_f, \alpha_g, \alpha_h)$ *is a global $(C \setminus \{0\})$-minimum point for $D_{\mathcal{V}}$.*

Proof Since x_0 is a global efficient point for P and since a constraint qualification condition holds, then for Theorem 1 $\exists \alpha_f \in C^+ \setminus \{0\}$, $\exists \alpha_g \in V^+$, $\exists \alpha_h \in \Re^p$, such that $\alpha_g^T g(x_0) = 0$ and

$$\nabla \mathcal{L}_{\mathcal{S}}(x_0, \alpha_f, \alpha_g, \alpha_h)^T (y - x_0) \leq 0 \quad \forall y \in X$$

This implies that $x_0 \in S_P$, $(x_0, \alpha_f, \alpha_g, \alpha_h)$ is feasible for $D_{\mathcal{V}}$, and $(1 - \delta)\alpha_g^T g(x_0) = 0$. The results then follows from Corollary 3 where $C^* = C \setminus \{0\}$ is assumed.   □

Like in Subsection 4.2, the following further duality result follows from the strong duality theorem.

Corollary 4 *Let us consider the primal problem P and the dual problem $D_{\mathcal{V}}$. Assume also that at least one of conditions (C_1) (with $C^* = C \setminus \{0\}$) and (C_2) is verified for all multipliers $(\alpha_f, \alpha_g, \alpha_h) \in (C^+ \times V^+ \times \Re^p)$, $\alpha_f \neq 0$, that X is convex and that a constraint qualification condition holds for problem P. Then, for all $x_1 \in S_P$ global efficient point for P and for all $(x_2, \alpha_f, \alpha_g, \alpha_h)$ global $(C \setminus \{0\})$-minimum point for $D_{\mathcal{V}}$ it results:*

$$f(x_1) - \mathcal{F}_{\mathcal{V}}(x_2, \alpha_f, \alpha_g, \alpha_h) \notin (C \cup -C) \setminus \{0\}$$

Proof For the strong duality theorem $\exists \widehat{\alpha}_f \in C^+ \setminus \{0\}$, $\exists \widehat{\alpha}_g \in V^+$, $\exists \widehat{\alpha}_h \in \Re^p$, such that $f(x_1) = \mathcal{F}_{\mathcal{V}}(x_1, \widehat{\alpha}_f, \widehat{\alpha}_g, \widehat{\alpha}_h)$ and $(x_1, \widehat{\alpha}_f, \widehat{\alpha}_g, \widehat{\alpha}_h)$ is a global $(C \setminus \{0\})$-minimum point for $D_{\mathcal{V}}$. Since $(x_2, \alpha_f, \alpha_g, \alpha_h)$ is a global $(C \setminus \{0\})$-minimum point for $D_{\mathcal{V}}$ too, it results:

$$\mathcal{F}_{\mathcal{V}}(x_1, \widehat{\alpha}_f, \widehat{\alpha}_g, \widehat{\alpha}_h) \notin \mathcal{F}_{\mathcal{V}}(x_2, \alpha_f, \alpha_g, \alpha_h) - C \setminus \{0\}$$
$$\mathcal{F}_{\mathcal{V}}(x_2, \alpha_f, \alpha_g, \alpha_h) \notin \mathcal{F}_{\mathcal{V}}(x_1, \widehat{\alpha}_f, \widehat{\alpha}_g, \widehat{\alpha}_h) - C \setminus \{0\}$$

As a consequence, it is:

$$\mathcal{F}_{\mathcal{V}}(x_1, \widehat{\alpha}_f, \widehat{\alpha}_g, \widehat{\alpha}_h) - \mathcal{F}_{\mathcal{V}}(x_2, \alpha_f, \alpha_g, \alpha_h) \notin (C \cup -C) \setminus \{0\}$$

and the thesis is proved being $f(x_1) = \mathcal{F}_{\mathcal{V}}(x_1, \widehat{\alpha}_f, \widehat{\alpha}_g, \widehat{\alpha}_h)$.   □

5   Duality in the case X is an open set

It is known that in the case X is an open set then the set constraint becomes meaningless and the maximum principle necessary optimality condition reduces to the classical Fritz John necessary optimality condition:

$$\nabla \mathcal{L}_\mathcal{S}(x_0, \alpha_f, \alpha_g, \alpha_h) = 0 \qquad (11)$$

This allows to propose the following further dual problems:

$$\overline{D}_\mathcal{S} : \begin{cases} \min \ \mathcal{F}_\mathcal{S}(x, \alpha_f, \alpha_g, \alpha_h) \\ \nabla \mathcal{L}_\mathcal{S}(x, \alpha_f, \alpha_g, \alpha_h) = 0 \\ (x, \alpha_f, \alpha_g, \alpha_h) \in S_D \end{cases} , \quad \overline{D}_\mathcal{V} : \begin{cases} C_- \min \ \mathcal{F}_\mathcal{V}(x, \alpha_f, \alpha_g, \alpha_h) \\ \nabla \mathcal{L}_\mathcal{S}(x, \alpha_f, \alpha_g, \alpha_h) = 0 \\ \alpha_f \neq 0 \\ (x, \alpha_f, \alpha_g, \alpha_h) \in S_D \end{cases}$$

where S_D has been already defined in Subsection 4.1.

Duality results concerning these two dual problems (a scalar and a vector one) will be based on the use of the semi-pseudoconcave properties introduced in Subsection 2.3. In this light, the provided results generalize the ones appeared in the literature based on generalized invexity properties.

5.1   Duality results for $\overline{D}_\mathcal{S}$

Theorem 9 (Weak duality for $\overline{D}_\mathcal{S}$) *Let us consider the primal problem P and the dual problem $\overline{D}_\mathcal{S}$. Assume also that function $\mathcal{L}_\mathcal{S}(x, \alpha_f, \alpha_g, \alpha_h)$ is semi-pseudoconcave with respect to variable x for all multipliers $(\alpha_f, \alpha_g, \alpha_h) \in (C^+ \times V^+ \times \Re^p)$. Then, $\forall x_1 \in S_P$ and $\forall (x_2, \alpha_f, \alpha_g, \alpha_h) \in S_D$ verifying (11) it is:*

$$\alpha_f^T f(x_1) \leq \mathcal{F}_\mathcal{S}(x_2, \alpha_f, \alpha_g, \alpha_h)$$

Proof Suppose by contradiction that $\exists x_1 \in S_P$ and $\exists (x_2, \alpha_f, \alpha_g, \alpha_h) \in S_D$ such that $\alpha_f^T f(x_1) > \mathcal{F}_\mathcal{S}(x_2, \alpha_f, \alpha_g, \alpha_h)$. Then, *ii)* of Lemma 2 yields:

$$\mathcal{L}_\mathcal{S}(x_1, \alpha_f, \alpha_g, \alpha_h) > \mathcal{L}_\mathcal{S}(x_2, \alpha_f, \alpha_g, \alpha_h)$$

and hence, from the semi-pseudoconcavity of $\mathcal{L}_\mathcal{S}(x, \alpha_f, \alpha_g, \alpha_h)$, it results:

$$\nabla \mathcal{L}_\mathcal{S}(x_2, \alpha_f, \alpha_g, \alpha_h) \neq 0$$

which contradicts (11).   □

Corollary 5 *Let us consider the primal problem P and the dual problem $\overline{D}_\mathcal{S}$. Assume also that function $\mathcal{L}_\mathcal{S}(x, \alpha_f, \alpha_g, \alpha_h)$ is semi-pseudoconcave with respect to variable x for all multipliers $(\alpha_f, \alpha_g, \alpha_h) \in (C^+ \times V^+ \times \Re^p)$. If $x_0 \in S_P$ and $(x_0, \alpha_f, \alpha_g, \alpha_h)$ is feasible for $\overline{D}_\mathcal{S}$, with $(1 - \delta)\alpha_g^T g(x_0) = 0$, then the following properties hold:*

i) $\alpha_f^T f(x_0) = \mathcal{F}_\mathcal{S}(x_0, \alpha_f, \alpha_g, \alpha_h)$;

ii) $(x_0, \alpha_f, \alpha_g, \alpha_h)$ *is a global minimum for* $\overline{D}_\mathcal{S}$;

iii) *if* $\alpha_f \neq 0$ *then* x_0 *is a global weak efficient point for* P.

Proof It is analogous to the one of Corollary 1. □

Theorem 10 (Strong duality for $\overline{\mathbf{D}}_\mathcal{S}$) *Let us consider the primal problem* P *and the dual problem* $\overline{D}_\mathcal{S}$. *Assume also that function* $\mathcal{L}_\mathcal{S}(x, \alpha_f, \alpha_g, \alpha_h)$ *is semi-pseudoconcave with respect to variable* x *for all multipliers* $(\alpha_f, \alpha_g, \alpha_h) \in (C^+ \times V^+ \times \Re^p)$ *and that* X *is open. Then, for all* $x_0 \in S_P$ *global efficient point for* P $\exists \alpha_f \in C^+$, $\exists \alpha_g \in V^+$, $\exists \alpha_h \in \Re^p$, $(\alpha_f, \alpha_g, \alpha_h) \neq 0$, *such that:*

i) $\alpha_f^T f(x_0) = \mathcal{F}_\mathcal{S}(x_0, \alpha_f, \alpha_g, \alpha_h)$;

ii) $(x_0, \alpha_f, \alpha_g, \alpha_h)$ *is a global minimum point for* $\overline{D}_\mathcal{S}$.

Proof The proof is analogous to the one of Theorem 6 and follows from Corollary 5. □

Corollary 6 *Let us consider the primal problem* P *and the dual problem* $\overline{D}_\mathcal{S}$. *Assume also that function* $\mathcal{L}_\mathcal{S}(x, \alpha_f, \alpha_g, \alpha_h)$ *is semi-pseudoconcave with respect to variable* x *for all multipliers* $(\alpha_f, \alpha_g, \alpha_h) \in (C^+ \times V^+ \times \Re^p)$ *and that* X *is open. Then, for all* $x_1 \in S_P$ *global efficient point for* P *and for all* $(x_2, \alpha_f, \alpha_g, \alpha_h)$ *global minimum point for* $\overline{D}_\mathcal{S}$ *it results:*

$$\exists \widehat{\alpha}_f \in C^+ \ such \ that \ \widehat{\alpha}_f^T f(x_1) = \mathcal{F}_\mathcal{S}(x_2, \alpha_f, \alpha_g, \alpha_h)$$

Proof The proof is analogous to the one of Corollary 2. □

5.2 Duality results for $\overline{\mathbf{D}}_\mathcal{V}$

Theorem 11 (Weak duality for $\overline{\mathbf{D}}_\mathcal{V}$) *Let us consider the primal problem* P *and the dual problem* $\overline{D}_\mathcal{V}$, *and let* C^* *be a cone such that* $Int(C) \subseteq C^* \subseteq C \setminus \{0\}$. *Assume also that at least one of the following conditions* (C_4) *and* (C_5) *is verified for all multipliers* $(\alpha_f, \alpha_g, \alpha_h) \in (C^+ \times V^+ \times \Re^p)$, $\alpha_f \neq 0$:

(C_4) *function* $\mathcal{L}_\mathcal{S}(x, \alpha_f, \alpha_g, \alpha_h)$ *is strictly semi-pseudoconcave with respect to* x;

(C_5) $C^* = Int(C)$ *and* $\mathcal{L}_\mathcal{S}(x, \alpha_f, \alpha_g, \alpha_h)$ *is semi-pseudoconcave with respect to* x.

Then, $\forall x_1 \in S_P$ *and* $\forall (x_2, \alpha_f, \alpha_g, \alpha_h) \in S_D$ *verifying (11),* $\alpha_f \neq 0$, *it is:*

$$f(x_1) \notin \mathcal{F}_\mathcal{V}(x_2, \alpha_f, \alpha_g, \alpha_h) + C^*$$

Proof Suppose by contradiction that $\exists x_1 \in S_P$, $\exists (x_2, \alpha_f, \alpha_g, \alpha_h) \in S_D$, $\alpha_f \neq 0$, such that

$$f(x_1) \in \mathcal{F}_\mathcal{V}(x_2, \alpha_f, \alpha_g, \alpha_h) + C^*$$

so that for *i)* of Lemma 3 it is $\mathcal{L}_\mathcal{V}(x_1, \alpha_f, \alpha_g, \alpha_h) \in \mathcal{L}_\mathcal{V}(x_2, \alpha_f, \alpha_g, \alpha_h) + C^*$.

Assume (C_4) holds, being $\alpha_f \neq 0$ it results $\mathcal{L}_\mathcal{S}(x_1, \alpha_f, \alpha_g, \alpha_h) \geq \mathcal{L}_\mathcal{S}(x_2, \alpha_f, \alpha_g, \alpha_h)$. Hence, the strict semi-pseudoconcavity of $\mathcal{L}_\mathcal{S}(x, \alpha_f, \alpha_g, \alpha_h)$ yields:

$$\nabla \mathcal{L}_\mathcal{S}(x_2, \alpha_f, \alpha_g, \alpha_h) \neq 0$$

which contradicts (11).

Assume (C_5) holds, being $\alpha_f \neq 0$ it results $\mathcal{L}_\mathcal{S}(x_1, \alpha_f, \alpha_g, \alpha_h) > \mathcal{L}_\mathcal{S}(x_2, \alpha_f, \alpha_g, \alpha_h)$. Hence, the semi-pseudoconcavity of $\mathcal{L}_\mathcal{S}(x, \alpha_f, \alpha_g, \alpha_h)$ yields:

$$\nabla \mathcal{L}_\mathcal{S}(x_2, \alpha_f, \alpha_g, \alpha_h) \neq 0$$

which again contradicts (11). $\qquad\square$

Corollary 7 *Let us consider the primal problem P and the dual problem $\overline{D}_\mathcal{V}$, and let C^* be a cone such that $Int(C) \subseteq C^* \subseteq C \setminus \{0\}$. Assume also that at least one of conditions (C_4) and (C_5) is verified for all multipliers $(\alpha_f, \alpha_g, \alpha_h) \in (C^+ \times V^+ \times \Re^p)$, $\alpha_f \neq 0$. If $x_0 \in S_P$ and $(x_0, \alpha_f, \alpha_g, \alpha_h)$ is feasible for $\overline{D}_\mathcal{V}$, with $(1-\delta)\alpha_g^T g(x_0) = 0$, then the following properties hold:*

   i) $f(x_0) = \mathcal{F}_\mathcal{V}(x_0, \alpha_f, \alpha_g, \alpha_h)$;

   ii) $(x_0, \alpha_f, \alpha_g, \alpha_h)$ is a global C^*-minimum point for $\overline{D}_\mathcal{V}$;

   iii) x_0 is a global C^*-efficient point for P.

   Proof It is analogous to the one of Corollary 3. $\qquad\square$

Theorem 12 (Strong duality for $\overline{D}_\mathcal{V}$) *Let us consider the primal problem P and the dual problem $\overline{D}_\mathcal{V}$. Assume also that condition (C_4) is verified for all multipliers $(\alpha_f, \alpha_g, \alpha_h) \in (C^+ \times V^+ \times \Re^p)$, $\alpha_f \neq 0$, that X is open and that a constraint qualification condition holds for problem P. Then, for all $x_0 \in S_P$ global efficient point for P $\exists \alpha_f \in C^+ \setminus \{0\}$, $\exists \alpha_g \in V^+$, $\exists \alpha_h \in \Re^p$, such that:*

   i) $f(x_0) = \mathcal{F}_\mathcal{V}(x_0, \alpha_f, \alpha_g, \alpha_h)$;

   ii) $(x_0, \alpha_f, \alpha_g, \alpha_h)$ is a global $(C \setminus \{0\})$-minimum point for $\overline{D}_\mathcal{V}$.

   Proof The proof is analogous to the one of Theorem 8 and follows from Corollary 7 where $C^* = C \setminus \{0\}$ is assumed. $\qquad\square$

Corollary 8 *Let us consider the primal problem* P *and the dual problem* $\overline{D}_{\mathcal{V}}$. *Assume also that condition* (C_4) *is verified for all multipliers* $(\alpha_f, \alpha_g, \alpha_h) \in (C^+ \times V^+ \times \Re^p)$, $\alpha_f \neq 0$, *that* X *is open and that a constraint qualification condition holds for problem* P. *Then, for all* $x_1 \in S_P$ *global efficient point for* P *and for all* $(x_2, \alpha_f, \alpha_g, \alpha_h)$ *global* $(C \setminus \{0\})$-*minimum point for* $\overline{D}_{\mathcal{V}}$ *it results:*

$$f(x_1) - \mathcal{F}_{\mathcal{V}}(x_2, \alpha_f, \alpha_g, \alpha_h) \notin (C \cup -C) \setminus \{0\}$$

Proof The proof is analogous to the one of Corollary 4. □

6   Conclusions

In this chapter we deal with a multiobjective problem where the image space of the objective function is ordered by an arbitrary convex, pointed cone and the feasible region is defined by equality constraint, inequality constraint and an arbitrary set which covers the constraints which cannot be expressed by means of neither equalities nor inequalities. The peculiarities of the primal problem suggest the introduction of vector and scalar dual problems which generalize the case where only equality and inequality constraints are considered. Along the line of the recent literature we introduce mixed-type dual problems where, according to the value of the parameters, both Wolfe-type and Mond-type dual problems are covered. The four presented dual problems can be summarized by means of the following table, which points out the used scalar or vector objective function and the considered necessary optimality condition:

	$\mathcal{L}_{\mathcal{S}}$	$\mathcal{L}_{\mathcal{D}}$
Max. principle conditions	$D_{\mathcal{S}}$	$D_{\mathcal{V}}$
K-K-T conditions	$\overline{D}_{\mathcal{S}}$	$\overline{D}_{\mathcal{V}}$

As we consider a primal problem with set constraint, efficient points verify maximum principle necessary optimality conditions and the feasible region of the dual problem is defined accordingly. On the other hand, when the primal problem has only inequality and equality constraints the Karush Kuhn Tucker (K-K-T) conditions are hidden in the definition of the feasible region of the dual problem. In each case, we define both a vector $(D_{\mathcal{V}}, \overline{D}_{\mathcal{V}})$ and a scalar $(D_{\mathcal{S}}, \overline{D}_{\mathcal{S}})$ dual problem and duality results are specified under suitable generalized concavity assumptions. When the feasible region is defined only by equality and inequality constraint, the introduction of a very easy generalized concavity properties allows us to recover, as special cases, results appeared in the recent literature.

References

[1] B. Aghezzaf and M. Hachimi, Generalized invexity and Duality in Multiobjective Programming Problems, Journal of Global Optimization 188(2000)91-101.

[2] B. Aghezzaf and M. Hachimi, Sufficiency and Duality in Multiobjective Programming Involving Generalized (F, ρ)−convexity, Journal of Mathematical Analysis and Applications 258(2001)617-628.

[3] T. Antczak, On G-invex multiobjective programming. Part II. Duality , Journal of Global Optimization 43(2009)111-140.

[4] M. Avriel, W.E. Diewert, S. Schaible and I. Zang, Generalized Concavity, Mathematical Concepts and Methods in Science and Engineering, Vol.36, Plenum Press, New York, 1998.

[5] D. Bhatia and P. Jain, Generalized (F, ρ)-convexity and duality for non smooth multiobjective programs, Optimization 31(1994)153-164.

[6] C.R. Bector, M.K. Bector, A. Gill and C. Singh, Duality for Vector Valued B-invex Programming, in *Generalized Convexity*, edited by S. Komlósi, T. Rapcsák and S. Schaible, Lecture Notes in Economics and Mathematical Systems, Vol. 405, Springer-Verlag, Berlin, (1994)358-373.

[7] C.R. Bector, Wolfe-Type Duality involving (B, η)-invex Functions for a Minmax Programming Problem, Journal of Mathematical Analysis and Application 201(1996)114-127.

[8] A. Cambini and L. Martein, Generalized Convexity and Optimization: Theory and Applications, Lecture Notes in Economics and Mathematical Systems, Vol. 616, Springer, Berlin, 2009.

[9] R. Cambini, Some new classes of generalized concave vector-valued functions, Optimization 36(1996)11-24.

[10] R. Cambini and L. Carosi, Duality in multiobjective optimization problems with set constraints, in Generalized Convexity, Generalized Monotonicity and Applications, edited by A. Eberhard, N. Hadjisavvas and D.T. Luc, Nonconvex optimization and its applications, vol.77, Springer, Berlin, (2005)131-146.

[11] R. Cambini and S. Komlósi, On the Scalarization of Pseudoconcavity and Pseudomonotonicity Concepts for Vector Valued Functions, in Generalized Convexity, Generalized Monotonicity: Recent Results, edited by J.-P. Crouzeix, J.-E. Martinez-Legaz and M. Volle, Nonconvex Optimization and Its Applications, Vol. 27, Kluwer Academic Publishers, Dordrecht, (1998)277-290.

[12] R. Cambini and S. Komlósi, On Polar Generalized Monotonicity in Vector Optimization, Optimization, Vol. 47(2000)111-121.

[13] R. Cambini and L. Martein, First and Second Order Characterizations of a Class of Pseudoconcave Vector Functions, in Generalized Convexity and Generalized Monotonicity, edited by N. Hadjisavvas, J.-E. Martinez-Legaz and J.P Penot, Lecture Notes in Economics and Mathematical Systems, Vol.502, Springer, Berlin, (2001)144-158.

[14] R. Cambini, Multiobjective problems with set constraints: from necessary optimality conditions to duality results, in Nonlinear analysis with applications in Economics, Energy and Transportation, A. Allevi, M. Bertocchi, A. Gnudi and I.V. Konnov (Eds.), Bergamo University Press, Bergamo (Ita), (2007)25-64.

[15] E. Caprari, ρ-invex functions and (F, ρ)-convex functions: properties and equivalencies, Optimization 52, No. 1(2003)65-74.

[16] A. Chinchuluun and M.P. Panos, A survey of recent developments in multiobjective optimization, Annals of Operations Research 154(2007)29-50

[17] R.R. Egudo, Efficiency and generalized convex duality for multiobjective programs, Journal of Mathematical Analysis and Applications 138(1989)84-94.

[18] R.R. Egudo and M.A. Hanson, Multiobjective duality with invexity, Journal of Mathematical Analisys and Applications 126(1987)469-477.

[19] G. Giorgi and A. Guerraggio, The notion of invexity in vector optimization: smooth and nonsmooth case, in Generalized Convexity, Generalized Monotonicity: Recent Results, edited by J.-P. Crouzeix, J.-E. Martinez-Legaz and M. Volle, Nonconvex Optimization and Its Applications, Vol. 27, Kluwer Academic Publishers, Dordrecht, (1998)389-405.

[20] G. Giorgi, B. Jiménez and V. Novo, Minimum principle-type optimality conditions for Pareto problems, International Journal of Pure and Applied Mathematics 10(2004)51-68.

[21] T.R. Gulati and M.A. Islam, Sufficiency and Duality in Multiobjective Programming Involving Generalized F-convexity, Journal of Mathematical Analysis and Applications 183(1994)181-195.

[22] M. Hachimi and B. Aghezzaf B., Journal of Mathematical Analisys and Application 319(2006)110-123

[23] M.A. Hanson and B. Mond, Necessary and Sufficiency Conditions in Constrained Optimization, Mathematical Programming 37(1987)51-58.

[24] V. Jeyakumar, ρ-Convexity and second order duality, Utilitas Mathematica 29(1986)71-85.

[25] B. Jiménez and V. Novo, A finite dimensional extension of Lyusternik theorem with applications to multiobjective optimization, Journal of Mathematical Analysis and Applications 270(2002)340-356.

[26] R.N. Kaul, S.K. Suneja and M.K. Srivastava, Optimality criteria and duality in multiple-objective optimization involving generalized invexity, Journal of Optimization Theory and Applications 80(1994)465-482.

[27] D.T. Luc, Theory of vector optimization, Lecture Notes in Economics and Mathematical Systems, Vol. 319, Springer-Verlag, Berlin, 1989.

[28] S.K. Mishra, On Sufficiency and Duality for Generalized Quasiconvex Nonsmooth Programs, Optimization 38(1996)223-235.

[29] B. Mond and T. Weir, Generalized concavity and duality, in Generalized Concavity and Duality in Optimization and Economics, edited by S. Schaible and W.T. Ziemba, Academic Press, New York, (1981)263-279.

[30] R.N. Mukherjee and Ch. Purnachandra Rao, Mixed Type Duality for Multiobjective Variational Problems, Journal of Mathematical Analysis and Applications 252(2000)571-586.

[31] S. Nobakhtian, Generalized (F, ρ)-Convexity and Duality in Nonsmooth Problems of Multiobjective Optimization, Journal of Optimization Theory and Applications 136(2008)61-68.

[32] V. Preda, On Efficiency and Duality in Multiobjective Programs, Journal of Mathematical Analysis and Applications 166(1992)365-377.

[33] N.G. Rueda, M.A. Hanson and C. Singh, Optimality and Duality with Generalized Convexity, Journal of Optimization Theory and Application 86(1995)491-500.

[34] Y. Sawaragi, H. Nakayama and T. Tanino, Theory of multiobjective optimization, Orlando: Academic, 1985.

[35] T. Tanino and Y. Sawaragi, Duality Theory in Multiobjective Programming, Journal of Optimization Theory and Application 27(1979)509-529.

[36] J.P. Vial, Strong convexity of sets and functions, Journal of Mathematical Economics 9(1982)187-205.

[37] T. Weir, Proper efficiency and duality for vector valued optimization problems, Journal of Australian Mathematical Society, (Series A) 43(1987)21-34.

[38] T. Weir, B. Mond and B.D. Craven, On duality for weakly minimized vector-valued optimization problems, Optimization 17(1986)711-721.

[39] P. Wolfe, A duality theorem for non-linear programming, Quarterly of Applied Mathematics 19(1961)239–244.

[40] Z. Xu, Mixed type duality in multiobjective programming problems, Journal of Mathematical Analysis and Applications 198(1996)621-635.

CHAPTER 9

Necessary and sufficient optimality conditions for continuous-time multiobjective optimization problems[*]

Adilson J. V. Brandão[†]   Valeriano Antunes de Oliveira [‡]

Marko Antonio Rojas-Medar[§]   Lucelina Batista Santos [¶]

Abstract

We discuss necessary and sufficient conditions of optimality for nonsmooth and smooth continuous-time multiobjective optimization problems under generalized convexity assumptions.

Keywords: multiobjective continuous-time programming, invexity, pseudoinvexity, optimality conditions, efficient solutions, generalized gradient.

1   Introduction

We will considere the continuous-time multiobjective programming problem, whose formulation is given below:

$$
\begin{array}{l}
\text{Minimize } \phi(x) := (\int_0^T f_1(t, x(t))dt, ..., \int_0^T f_p(t, x(t))dt) \\
\text{subject to: } g_i(t, x(t)) \le 0 \text{ a.e. } t \in [0, T], \\
i \in I = \{1, ..., m\}, \ x \in X.
\end{array}
\qquad \text{(VCNP)}
$$

Here, X is open, nonempty convex subset of the Banach space $L_\infty^n[0, T]$ of all n-dimensional vector-valued Lebesgue measurable functions which are essentially bounded, defined on the compact interval $[0, T]$ with the norm $|| \cdot ||_\infty$ defined by

$$
||x||_\infty = \max_{k=1,...,n} \text{ess sup } |x_k(t)|, t \in [0, T]\},
$$

[*]This work was partially supported by the grant MTM2007-063432 of the Science and Education Spanish Ministry, CNpq-Brazil and FONDECYT- Chile

[†]Departamento de Matemática. Universidade Federal de São Carlos, Campus Sorocaba. Rodovia João Leme dos Santos, Km 110 - SP-264 Itinga. 18052-780 - Sorocaba, SP - Brasil. e-mail: adilsonvb@ufscar.br manuel.arana@uca.es

[‡]Universidade Federal de Uberlândia, Faculdade de Ciências Integradas do Pontal. Av. José João Dib, 2545 Progresso 38302-000 - Ituiutaba, MG - Brasil. e-mail: valeriano@pontal.ufu.br

[§]Universidad Del Bío-Bío. Facultad de Ciencias. Departamento de Ciencias Básicas. Casilla: 447 Av. Andrés Bello S/N Chillán- Chile. e-mail: marko@ueubiobio.cl

[¶]Departamento de Matemática. Universidade Federal do Paraná. CP 19081 CEP, 81531-990 Curitiba, Paraná, Brasil e-mail: lucelina@ufpr.br

where for each $t \in [0, T]$, $x_k(t)$ is the k-th component of $x(t) \in \mathbb{R}^n$, ϕ is a real-valued function defined on X, $g_i(t, x(t)) = \gamma_i(x)(t)$ and $f_j(t, x(t)) = \Gamma_j(x)(t)$, where γ_i, $i \in I$ are maps from X into the normed space $\Lambda_1^m[0, T]$ of all Lebesgue measurable essentially bounded m-dimensional vector functions defined on $[0, T]$ with the norm $|| \cdot ||_1$ defined by

$$||y||_1 = \max_{j=1,\dots,m} \int_0^T |y_j(t)| dt$$

and Γ_j, $j \in J = \{1, ..., m\}$ are maps from X into the normed space $\Lambda_1^1[0, T]$.

The mono-objective version of this problem was introduced by Bellman in 1953 [2] in connection with production-inventory called "bottleneck problems". He considered a type of optimization problems which is known as continuous-time linear programming, formulated its dual and provide some duality relations. He also suggested some computational procedures.

Since then, a lot of authors have extended his theory to wider classes of continuous-time problems. See [3], [13]-[16].

In such papers, the authors study the mono-objective case, but in many applications, it is necessary to minimize many objectives and so, the multiobjective is more general and suitable for some applications.

Our main of this work is to state necessary and sufficient conditions for (VCNP). This is accomplished through generalizations of the differentiable versions of Fritz-John and Karush-Kuhn-Tucker theorems in [15] to the Lipschitz case. Also, we will consider the particular case where the functions of the problems are differentiable. For this problem, we introduce the notion of Karush-Kuhn-Tucker pseudoinvexity and using this concept we show that Karush-Kuhn-tucker pseuoinvexity is a necessary and sufficient condition for a vector Karush-Kuhn-Tucker point to be a weakly efficient solution. We organized this Chapter into four sections: in Section 2, we give some preliminaries; in Sections 3 and 4 we obtain, respectively, necessary and sufficient conditions for weak efficiency for the nonsmooth problem (VCNP) and, finally, in Section 5, we discuss the particular case where the problem is differentiable.

2 Preliminaries

In this Section, we fix some basic concepts and notation adhered to this Chapter.

2.1 Support functions and integration of multifunctions

Let B a Banach space. We recall that the support function of a nonempty closed convex subset D of B is the unction $\sigma_D : B^* \to \mathbb{R} \cup \{+\infty\}$ defined by

$$\sigma_D(\xi) = \sup_{x \in D} \langle \xi, x \rangle,$$

where B^* is the topological dual of B, equipped with the weak*-topology and $\langle \cdot, \cdot \rangle$ is the canonical bilinear form between B^* and B. We now state some basic known results of support functions which are needed in the sequel.

Theorem 1 *[Hormander] Let C and D be nonempty closed convex subsets of B and suppose that Σ and Δ are noenempty weak*-closed convex subsets of B^*. Also, let $\mu, \lambda \geq 0$ be given scalars. Then,*

$$\mu \sigma_C(\xi) + \lambda \sigma_D(\xi) = \sigma_{\mu C + \lambda D}(\xi),$$
$$\mu \sigma_\Delta(x) + \lambda \sigma_\Sigma(x) = \sigma_{\mu \Delta + \lambda \Sigma}(x)$$

for every $\xi \in B^*$ and $x \in B$.

Given a multifunction (that is, a point-to-set map) $\Gamma : [0, T] \to \mathbb{R}^n$, denote by $S^1[0, T]$ the following set

$$S^1(\Gamma) = \{f \in L_1^n[0, T]; f(t) \in \Gamma(t) \text{ a.e. } t \in [0, T]\}.$$

We define the *integral* of Γ, denoted by $\int_0^T \Gamma(t)dt$, as the following subset of $\mathbb{R}^n$:

$$\int_0^T \Gamma(t)dt := \{\int_0^T f(t)dt : f \in S^1(\Gamma)\}.$$

We say that the multifunction $\Gamma : [0, T] \to \mathbb{R}^n$ is said to be measurable if for every open set $U \subset \mathbb{R}^n$, the set $\Gamma^{-1}(U) = \{t \in [0, T] : \Gamma(t) \cap U \neq \emptyset\}$ is Lebesgue-measurable.

A multifunction Γ is said to be *integrably bounded* if Γ is measurable and there exists a integrable function $z : [0, T] \to \mathbb{R}_+$, such that

$$||\Gamma(t)|| \leq z(t) \text{ a.e. on } [0, T].$$

Theorem 2 *If Γ is a integrably bounded multifunction taking compact values of $\mathbb{R}^n$, then*

$$\sigma_{\int_0^T \Gamma(t)dt}(v) = \int_0^T \sigma_{\Gamma(t)}(v)dt, \forall v \in \mathbb{R}^n.$$

For more details, see [1].

2.2 Generalized gradients and derivatives

Let Z be a Banach space and $\psi : Z \to r$ is a locally Lipschitz function, i. e., for each $x \in Z$, there exists $\varepsilon > 0$ and a constant $K > 0$ depending on ε, such that

$$|\psi(x_1) - \psi(x_2)| \leq K||x_1 - x_2||, \forall x_1, x_2 \in x + \varepsilon \mathcal{B},$$

where $\mathcal{B}$ is the open unit ball of Z.

The *Clarke generalized directional derivative* of ψ at x in the direction of a given $v \in Z$, denoted by $\psi^0(x; v)$ is defined by

$$\psi^0(x; v) := \limsup_{\substack{y \to x \\ s \to 0^+}} \frac{\psi(y + sv) - \psi(y)}{s}.$$

The *generalized gradient* of ψ at x, denoted by $\partial\psi(x)$, is defined by

$$\partial\psi(x) := \{\xi \in Z^* : \langle \xi, v \rangle \leq \psi^0(x; v), \forall v \in Z\}.$$

It can be proved that $\psi^0(x; \cdot) = \sigma_{\partial f(x)}$.

We say that ψ is *Clarke regular* at $x \in U$ if for all $v \in Z$, the usual one-sided directional derivative of ψ at x in the direction $v \in Z$, denoted by $\psi'(x; v)$ exists and $\psi'(x; v) = \psi^0(x; v)$.

We recall that if $\psi_i(\cdot)$ are Clarke regular at x, for $i = 1, ..., n$ and $\lambda_i > 0$, then

$$\partial\left(\sum_{i=1}^{n} \lambda_i \psi_i(x)\right) = \sum_{i=1}^{n} \lambda_i \partial\psi_i(x).$$

(in general, we can guarantee only the inclusion $\partial(\sum_{i=1}^{n} \lambda_i \psi_i(x)) \subset \sum_{i=1}^{n} \lambda_i \partial\psi_i(x)$).

Let $\mathbb{F}$ be the set of all feasible solutions of (VCNP) (which we will suppose nonempty), i.e.,

$$\mathbb{F} = \{x \in X : g_i(t, x(t)) \leq 0, \text{ a.e. in } [0, T], \ i \in I\}.$$

Let V be an open convex subset of $\mathbb{R}^n$ containing the set $\{x(t) \in \mathbb{R}^n : x \in \mathbb{F}, t \in [0, T]\}$.

We assume that $f_j, j \in J$ and $g_i, i \in I$ are real functions defined on $[0, T] \times V$. The functions $t \mapsto f_j(t, x(t)), j \in J$ are assumed to be Lebesgue measurable and integrable for all $x \in X$.

Also we assume that for every given $a \in V$, there exist $\varepsilon > 0$ and $k > 0$ such that for all $t \in [0, T]$ and every $x_1, x_2 \in a + \varepsilon B$, we have

$$|f_j(t, x_1) - f(t, x_2)| \leq k\|x_1 - x_2\|, \ \forall j \in J$$

(where B denotes the unit ball of $\mathbb{R}^n$).

Similar hypotheses are assumed for $g_i, i \in I$. Hence, $f_j(t, \cdot), j \in J$ and $g_i(t, \cdot), i \in I$ are locally Lipshitz on V throughout $[0, T]$.

Without lost of generality, we can suppose that the Lipschitz constant is the same for all involved functions.

Now, assume $\overline{x} \in X$ and $h \in L_\infty^n[0, T]$ are given. The continuous-time Clarke generalized directional derivatives of f_j and g_i are given by:

$$f_j^0(t, \overline{x}(t); h(t)) := \Gamma_j^0(\overline{x}; h)(t) := \limsup_{\substack{y \to \overline{x} \\ s \to 0^+}} \frac{\Gamma_j(y + sh)(t) - \Gamma_j(y)(t)}{s}$$

and

$$g_i^0(t, \overline{x}(t); h(t)) := \gamma_i^0(\overline{x}; h)(t) := \lim_{\substack{y \to \overline{x} \\ s \to 0^+}} \sup \frac{\gamma_i(y + sh)(t) - \gamma_i(y)(t)}{s}$$

a.e. in $[0, T]$.

it follows easily from the assumptions one that the functions

$$t \;\mapsto\; f_j^0(t, \overline{x}(t); h(t)), \; j \in J$$
$$t \;\mapsto\; g_i^0(t, \overline{x}(t); h(t)), \; i \in I$$

are Lebesgue measurable and integrable for all $\overline{x} \in X$ and $h \in L_\infty^n[0, T]$.

2.3 Generalized convexity

The concept of invex function was introduced by Hanson in [7] and it was extended for nonsmooth functions in [5], [12]. Let U be a nonempty subset of Z and $\psi : U \to \mathbb{R}$ be a locally Lipshitz function on U. The function ψ is said to be invex at $\overline{z} \in U$ over U if there exists a function $\eta : U \times U \to Z$ such that

$$\psi(z) - \psi(\overline{z}) \geq \psi^0(\overline{z}; \eta(z, \overline{z}))$$

for all $z \in U$. We say that ψ is strictly invex at $\overline{z}$ over U, if the above inequaliy is strict for every $z \neq \overline{z}$.

We also can define an invexity notion in the continuous-time context [13].

Definition 3 *Let $U \subset \mathbb{R}^n$ be a nonempty subset of $\mathbb{R}^n$ and $\overline{x} \in X$. Suppose that $\psi : [0, T] \times U \to \mathbb{R}$ is locally Lipschitz function throughout $[0, T]$. The function $\psi(t, \cdot)$ is said to be invex at $\overline{x}(t)$ over U if there exists $\eta : U \times U \to \mathbb{R}^n$, such that the function $t \mapsto \eta(x(t), \overline{x}(t))$ is in $L_\infty^n[0, T]$ and*

$$\psi(t, x(t)) - \psi(t, \overline{x}(t)) \geq \psi^0(t, \overline{x}(t); \eta(x(t), \overline{x}(t)) \; a.e. \; in \; [0, T]$$

for all $x \in X$. We say that ψ is strictly invex at $\overline{x}(t)$ over U if the above inequality is strict for every $x(t) \neq \overline{x}(t)$ a.e. in $[0, T]$.

3 Necessary conditions: nonsmooth case

Let $\phi_j : X \to \mathbb{R}$ be a function defined by

$$\phi_j(x) = \int_0^T f_j(t, x(t)) dt.$$

We say that $\overline{x} \in X$ is a *local (global) efficient solution* of (VCNP) if there exists a neighborhood U of $\overline{x}$ such that does not exist $x \in \mathbb{F} \cap U$ such that $\phi_j(x) \leq \phi_j(\overline{x})$, for every

$j \in J$, with strict inequality holding for some j (respectively: if does not exist $x \in \mathbb{F}$ such that $\phi_j(x) \leq \phi_j(\overline{x})$, for every $j \in J$, with strict inequality holding for some j).

A point $\overline{x} \in X$ is said a *local (global) weakly efficient solution* of (VCNP) if there exists a neighborhood U of $\overline{x}$ such that does not exist $x \in \mathbb{F} \cap U$ such that $\phi_j(x) < \phi_j(\overline{x})$, for every $j \in J$ (respectively: if does not exist $x \in \mathbb{F}$ such that $\phi_j(x) < \phi_j(\overline{x})$, for every $j \in J$).

Geoffrion [6] introduced the concept of proper efficiency which eliminates efficient points of a certain anomalous type: $\overline{x}$ is said to be a *properly efficient solution* of (VCNP) if it is efficient and if there exists a scalar $M > 0$ such that for each i, we have

$$\frac{\phi_i(\overline{x}) - \phi_i(x)}{\phi_j(x) - \phi_j(\overline{x})} \leq M$$

for some j such that $\phi_j(x) > \phi_j(\overline{x})$ and $\phi_i(x) < \phi_i(\overline{x})$.

Consider the following cones with vertex at 0 in $L_\infty^n[0, T]$:

$$\mathcal{K}(\phi_j; \overline{x}) \quad : \quad = \{h \in L_\infty^n[0, T] : \phi_j^0(\overline{x}; h) < 0\}, \; j \in J;$$

$$\mathcal{K}(g_i; \overline{x}) \quad : \quad = \{h \in L_\infty^n[0, T] : g_i^0(t, \overline{x}(t); h(t)) < 0 \text{ a.e. } t \in A_i(\overline{x})\}, \; i \in I$$

where $A_i(\overline{x}) = \{t \in [0, T]; g_i(t, \overline{x}(t)) = 0\}$.

We are now in position to provide a geometric characterization of local weak efficiency for (VCNP).

Theorem 4 *Let $\overline{x}$ be a local weak efficient solution of (VCNP). Then*

$$\bigcap_{i \in I} \mathcal{K}(g_i; \overline{x}) \cap \bigcap_{j \in J} \mathcal{K}(g_i; \overline{x}) = \emptyset. \tag{1}$$

Proof. Suppose that the intersection of cones (1) is nonempty and take $h \in L_\infty^n[0, T]$ in this intersection. It follows from $\limsup$ properties and the continuity of the functions that there is a real number $\delta > 0$ such that for every $\lambda \in (0, \delta)$, $\overline{x} + \lambda h \in X$ (because X is open) and

$$g_i(t, \overline{x}(t) + \lambda h(t)) \quad \leq \quad 0 \text{ a.e. } t \in [0, T], i \in I$$
$$\phi_j(\overline{x} + \lambda h) \quad < \quad \phi_j(\overline{x}), \; j \in J.$$

But it means that $\overline{x} + \lambda h$ is a feasible solution of (VCNP), for every $\lambda \in (0, \delta)$ with objective value better than $\overline{x}$. This contradicts the local efficiency of $\overline{x}$. Therefore, the intersection (1) is empty. ∎

In the following, we estate a transposition theorem, known as Generalized Gordan's Theorem and it is a generalization of a similar result proved by Zalmai [16]. It is the key to move from the geometric optimality condition obtained above to the main results on first order necessary optimality conditions.

For the next result, the domain of definition of the spaces $L_\infty^n[0, T]$, $L_\infty^m[0, T]$, $\Lambda_1^m[0, T]$ is replaced with a nonzero Lebesgue measure set $A \subset [0, T]$.

Theorem 5 *Let $A \subset [0,T]$ be a set of positive Lebesgue measure and X be a nonmpty convex subset of $L_\infty^n(A)$ and $p_i : V \times A \to \mathbb{R}$, $i \in I = \{1, ..., m\}$ be defined by $p_i(t, x(t)) = \pi_i(x)(t)$, where V is an open subset of $\mathbb{R}^n$, $\pi = (\pi_1, ..., \pi_m)$ is a map from X to $\Lambda_1^m(A)$ and suppose that p_i is convex with respect to its argument on V throughout A. Then, exactly one of the following systems is consistent:*

1. *There is $x \in X$ such that $p_i(t, x(t)) < 0$ a.e. $t \in A$, $i \in I$;*

2. *There is a nonzero m-vector function $u \in L_\infty^m(A)$, $u_i(t) \geq 0$ a.e. $t \in A$, $i \in I$, such that $\int_0^T \sum_{i \in I} u_i(t) p_i(t, x(t)) dt \geq 0$, for all $x \in X$.*

Proof. The proof of this Theorem is very similar with the proof of Theorem 3.2 in [16], replacing $[0, T]$ by A. ■

We are in position to derive a continuous-time analogue of the Fritz-John necessary optimality conditions translating the geometric optimality conditions into algebraic statements. This is made through the use of the Generalized Gordan´s Theorem. We also point out that the Fritz-John necessary conditions generalizes the nonsmooth mono-objective case studied in [3].

Theorem 6 *Let $\overline{x} \in \mathbb{F}$ be a feasible point of (VCNP) and suppose that $f_j(t, \cdot)$, $j \in J$ and $g_i(t, \cdot)$, $i \in I$ be Lipschitz functions near $\overline{x}(t)$. If $\overline{x}$ is a local weak efficient solution of (VCNP) then there exist $\overline{\mu}_j \in \mathbb{R}$, $j \in J$, $\overline{\lambda}_i \in L_\infty^m[0,T]$, $i \in I$ such that*

$$0 \in \int_0^T [\sum_{j \in J} \overline{\mu}_j \partial_x(f_j(t, \overline{x}(t))) + \sum_{i \in I} \overline{\lambda}_i \partial_x g_i(t, \overline{x}(t))] dt; \tag{2}$$

$$\overline{\mu}_j \geq 0, \; j \in J, \; \overline{\lambda}_i(t) \geq 0 \text{ a.e. } t \in [0,T], i \in I; \tag{3}$$

$$(\overline{\mu}, \overline{\lambda}(t)) \neq 0 \text{ a.e. } t \in [0,T]; \tag{4}$$

$$\overline{\lambda}_i(t) g_i(t, \overline{x}(t)) = 0 \text{ a.e. } t \in [0,T], i \in I. \tag{5}$$

For to prove Theorem 6, we shall proceed under the *Interim Hypothesis*: We suppose that (VCNP) has only one constraint

$$g(t, x(t)) \leq 0 \text{ a.e. in } [0,T].$$

The removal of this interim hypothesis will be done at the end of the proof.

We denote:

$$A(\overline{x}) = \{t \in [0,T] : g(t, \overline{x}(t)) = 0\};$$
$$\mathcal{K}(g, \overline{x}) = \{h \in L_\infty^n[0,T]; g^0(t, \overline{x}(t); h(t)) < 0, t \in A(\overline{x})\}.$$

Lemma 7 *Suppose that (VCNP) satisfies the Interim Hypothesis. Let $\overline{x} \in \mathbb{F}$ be a feasible point of (VCNP) and suppose that that $f_j(t, \cdot), j \in J$ and $g(t, \cdot)$ are Lipschitz near $\overline{x}(t)$ throughout $[0, T]$. If $\overline{x}$ is a local weak efficient solution of (VCNP) then there exist $\overline{\mu} \in \mathbb{R}^p, \overline{\lambda} \in L_{\infty}^n[0, T]$ such that*

$$0 \leq \int_0^T [\sum_{j \in J} \overline{\mu}_j f_j^0(t, \overline{x}(t); h(t)) + \overline{\lambda}(t) g^0(t, \overline{x}(t); h(t))] dt, \forall h \in L_{\infty}^n[0, T]; \tag{6}$$

$$\overline{\mu}_j \geq 0, j \in J, \overline{\lambda}(t) \geq 0 \ a.e. \ t \in [0, T]; \tag{7}$$

$$(\overline{\mu}, \overline{\lambda}(t)) \neq 0 \ a.e. \ t \in [0, T]; \tag{8}$$

$$\overline{\lambda}(t) g(t, \overline{x}(t)) = 0 \ a.e. \ t \in [0, T]. \tag{9}$$

Proof (of Lemma 7): If $\overline{x}$ is a local weak efficient solution of (VCNP), then by Theorem 4,

$$\mathcal{K}(g, \overline{x}) \cap \bigcap_{j \in J} \mathcal{K}(\phi_j, \overline{x}) = \emptyset.$$

Hence, there is no $h \in L_{\infty}^n[0, T]$ such that

$$\phi_j^0(\overline{x}; h) < 0, j \in J;$$
$$g^0(t, \overline{x}(t); h(t)) < 0, \text{a.e.} \ t \in A(\overline{x}).$$

By making use of Theorem 5, that there are $\mu \in L_{\infty}^p[0, T], \lambda \in L^1[0, T]$ with $\mu_j(t) \geq 0$ and $\lambda(t) \geq 0$ a.e. $t \in [0, T]$ not all identically zero and such that

$$0 \leq \int_{A(\overline{x})} [\sum_{j \in J} \mu_j(t) \phi_j^0(\overline{x}; h) + \lambda(t) g^0(t, \overline{x}(t); h(t))] dt, \forall h \in L_{\infty}^n[0, T].$$

Setting: $\overline{\mu}_j = \int_{A(\overline{x})} \mu_j(t) dt$ and $\overline{\lambda}(t) = \begin{cases} \lambda(t), t \in A(\overline{x}) \\ 0, \text{ otherwise} \end{cases}$, we obtain

$$\begin{aligned} 0 &\leq \int_{A(\overline{x})} [\sum_{j \in J} \mu_j(t) \phi_j^0(\overline{x}; h) + \lambda(t) g^0(t, \overline{x}(t); h(t))] dt \\ &\leq \int_0^T [\mu_j f_j^0(t, \overline{x}(t); h(t)) + \overline{\lambda}(t) g^0(t, \overline{x}(t); h(t))] dt \end{aligned}$$

for all $h \in L_{\infty}^n[0, T]$ (The Fatou's Lemma is used in the last inequality). Thus, (6) is proved. The remaining assertions of the Lemma 7 follow immediately.∎

For to conclude the proof of the Theorem 6 under interim hypothesis, we can prove that

$$0 \in \int_0^T [\sum_{j \in J} \overline{\mu}_j \partial_x f_j(t, \overline{x}(t)) + \overline{\lambda}(t) \partial g_x(t, \overline{x}(t))] dt. \tag{10}$$

Note that (10) can be rewritten in terms of the support unctions, as follows:

$$0 \le \int_0^T [\sum_{j \in J} \overline{\mu}_j \sigma_{\partial_x f_j(t, \overline{x}(t))}(h(t)) + \overline{\lambda}(t) \sigma_{\partial g_x(t, \overline{x}(t))}(h(t)) dt, \forall h \in L_\infty^n[0, T].$$

Since the above inequality holds for all $h \in L_\infty^n[0, T]$, it holds, in particular, for constant functions $h(t) = v \in \mathbb{R}^n$.

It can be easily verified that the multifunction

$$t \mapsto \sum_{j \in J} \overline{\mu}_j \partial_x f_j(t, \overline{x}(t)) + \overline{\lambda}(t) \partial_x g(t, \overline{x}(t))$$

is integrably bounded and takes values compact subsets of $\mathbb{R}^n$. By Theorem 2, we have

$$
\begin{aligned}
0 &\le \int_0^T [\sum_{j \in J} \overline{\mu}_j \sigma_{\partial_x f_j(t, \overline{x}(t))}(v) + \overline{\lambda}(t) \sigma_{\partial g_x(t, \overline{x}(t))}(v) dt \\
&= \int_0^T [\sigma_{\overline{\mu}_j \partial_x f_j(t, \overline{x}(t)) + \overline{\lambda}(t) \partial_x g(t, \overline{x}(t))}](v) dt \\
&= \sigma_{\int_0^T [\overline{\mu}_j \partial_x f_j(t, \overline{x}(t)) + \overline{\lambda}(t) \partial_x g(t, \overline{x}(t))] dt}(v)
\end{aligned}
$$

and it is equivalent to (10), which finishes the proof of the Theorem 6, under the interim hypothesis.

For to conclude the proof of the Theorem 6, it remains to removal of the interim hypothesis.

Removal of the interim hypothesis: Suppose that (VCNP) has m constraints $g_i(t, x(t)) \le 0$ a.e. in $[0, T]$ and $\overline{x}$ is a local weak efficient solution of (VCNP). We can reduce the m constraints of (VCNP) to just one by defining $g(t, x(t)) = \max_{i=1,\dots,m} g_i(t, x(t)) \le 0$ a.e. in $[0, T]$. The point $\overline{x}$ is also an optimal solution of the modified problem. Let $I(t, x) := \{i \in I : g_i(t, x(t)) = g(t, x(t))\}$. From what that we have been proved under the interim hypothesis, there exist $\overline{\mu} \in \mathbb{R}^p$, $\lambda \in L_\infty^m[0, T]$, satisfying

$$0 \in \int_0^T [\sum_{j \in J} \overline{\mu}_j \partial_x f_j(t, \overline{x}(t)) + \lambda(t) \partial_x g(t, \overline{x}(t))] dt \qquad (11)$$

and (7)-(9). It can be deduced from (11) and the definition of integration of multifunctions that there exists a measurable function $e(t) \in \partial_x g(t, \overline{x}(t))$ a.e. such that

$$0 \in \int_0^T [\sum_{j \in J} \overline{\mu}_j \partial_x f_j(t, \overline{x}(t)) + \lambda(t) e(t)] dt. \qquad (12)$$

We have the following Lemma:

Lemma 8 *There exists $v \in L_\infty^m[0,T]$, $v \geq 0$ a.e. satisfying:*

1. $v_i(t) = 0$ *whenever* $g_i(t, \overline{x}(t)) \neq g(t, \overline{x}(t))$, $i = 1, ..., m$;

2. $\sum_{i=1}^m v_i(t) = 1$ *a.e. in* $[0,T]$;

3. $e(t) \subset \sum_{i=1}^m v_i(t)\partial_x g_i(t, \overline{x}(t))$ *a.e. in* $[0,T]$.

Proof. (of Lemma 8) For each t where $\partial_x(t, \overline{x}(t))$ is well defined it can be proved that

$$\partial_x g(t, \overline{x}(t)) \subset \mathrm{co}\{\partial_x g_i(t, \overline{x}(t)) : i \in I(t, \overline{x}(t))\}.$$

See [4]. Since $e(t) \in \partial_x g(t, \overline{x}(t))$ a.e. in $[0,T]$, we obtain

$$e(t) \in \mathrm{co}\{\partial_x g_i(t, \overline{x}(t)) : i \in I(t, \overline{x}(t))\}.$$

Now, define

$$V(t) \quad : \quad = \{(v_1, ..., v_m) \in \mathbb{R}^m : \sum_{i=1}^m v_i = 1, v_i \geq 0, \ v = 0 \text{ if } g_i(t, \overline{x}(t)) < g(t, \overline{x}(t)),$$

$$e(t) \quad \in \quad \sum_{i=1}^m v_i \partial_x g_i(t, \overline{x}(t))\}.$$

The set $V(t)$ is obviously nonempty and closed a.e. in $[0,T]$. It follows from standard measurable selection theorems (see e. g. [1]) that we can choose measurable functions $v_1(t), ..., v_m(t)$ defined on $[0,T]$ such that $(v_1(t), ..., v_m(t)) \in V(t)$ a.e. in $[0,T]$. The proof of the Lemma 8 follows immediately. ∎

Now defining $\overline{\lambda}_i(t) := \lambda(t)v_i(t)$, it follows easily from Lemma 8 and (12) that assertions (2)-(5) of Theorem 6 are valid and we conclude the proof of Theorem 6.∎

Remark 9 *In Theorem 6, if the functions $f_j(t, \cdot)$, $j \in J$ and $g_i(t, \cdot)$, $i \in I$ are Clarke regular, then the condition (2) is equivalent to*

$$0 \in \partial_x L(\overline{x}, \overline{\mu}, \overline{\lambda}),$$

where $L(x, \mu, \lambda)$ is the Lagrangean function defined by

$$L(x, \mu, \lambda) := \int_0^T [\sum_{j \in J} \mu_j f_j(t, x(t)) + \sum_{i \in I} \lambda_i(t) g_i(t, x(t))] dt$$

with $x \in X, \mu \in \mathbb{R}^p, \lambda \in L_\infty^m[0,T]$.

In the necessary conditions given by Theorem 6, there is no guarantee that the Lagrange multiplier $\overline{\mu}$ associated with the objective function will be nonzero. It is usual to assume some kind of regularity condition on the restrictions of the problem, to make sure that this multiplier is nonzero. These regularity conditions are usually refereed to as constraint qualifications. We will assume the following natural constraint qualification:

$$\bigcap_{i \in I} \mathcal{K}(g_i, \overline{x}) \neq \emptyset. \tag{CQ}$$

Theorem 10 *Let $\overline{x} \in \mathbb{F}$ and suppose that $f_j(t, \cdot)$, $j \in J$ and $g_i(t, \cdot)$, $i \in I$ are Lipschitz near $\overline{x}(t)$. Let $\overline{x}$ be a local weakly efficient solution of (VCNP) and also, assume that (CQ) holds on $\overline{x}$. Then, there exist $\overline{\mu}_j \in \mathbb{R}$, $j \in J$ and $\overline{\lambda}_i \in L_\infty^m[0, T]$, $i \in I$ such that*

$$0 \in \int_0^T [\sum_{j \in J} \overline{\mu}_j \partial_x(f_j(t, \overline{x}(t))) + \sum_{i \in I} \overline{\lambda}_i \partial_x g_i(t, \overline{x}(t))] dt; \tag{13}$$

$$\overline{\mu}_j \geq 0, \ j \in J, \ \overline{\lambda}_i(t) \geq 0 \ a.e. \ t \in [0, T], i \in I; \tag{14}$$

$$\mu \neq 0; \tag{15}$$

$$\overline{\lambda}_i(t) g_i(t, \overline{x}(t)) = 0 \ a.e. \ t \in [0, T], i \in I. \tag{16}$$

Proof. Firstly, we will prove the Theorem 10 under the interim hypothesis, in the similar fashion that we proved the previous theorem. We will suppose that (VCNP) has only one constraint $g(t, x(t)) \leq 0$ a.e. in $[0, T]$. If $\overline{x}$ is a local weak efficient solution of (VCNP) then there exist $\overline{\mu} \in \mathbb{R}^p$, $\overline{\lambda} \in L_\infty^1[0, T]$ such that (6)-(9) hold true. If $\overline{\mu} = 0$ in (6) would reduce to

$$0 \leq \int_0^T \overline{\lambda}(t) g^0(t, \overline{x}(t); h(t)) dt, \forall h \in L_\infty^n[0, T].$$

Hence, by the Generalized Gordan´s Lemma, there is no $h \in X$ such that

$$g^0(t, \overline{x}(t); h(t)) < 0 \ \text{a.e. in} \ [0, T],$$

that contradicts the constraint qualification (CQ). So, $\overline{\mu} \neq 0$ and the theorem follows from the inequality

$$0 \leq \int_0^T [\sum_{j \in J} \overline{\mu}_j f_j^0(t, \overline{x}(t); h(t)) + \overline{\lambda}(t) g^0(t, \overline{x}(t); h(t))] dt, \forall h \in L_\infty^n[0, T]$$

by using similar arguments to those used in the proof of condition (2) of Theorem 6.

Removal of the Interim Hypothesis: Let $g(t, x(t)) = \max_{i \in I} g_i(t, x(t))$. Note that the constraint qualification (CQ) implies $K(g, \overline{x}) \neq \emptyset$ (see Brandão *et al.* [3]) and similar arguments to those employed in the proof of Theorem 6 (in the removal of interim hypothesis) yields the desired result. ■

4 Sufficient conditions: nonsmooth case

In this Section we obtain sufficient conditions for (weak) efficiency for (VCNP) by using of suitable hypotheses of generalized convexity.

Theorem 11 *[Fritz-John sufficiency] Let $\overline{x} \in \mathbb{F}$ and suppose that $f_j(t, \cdot)$, $j \in J$ are invex at $\overline{x}(t)$ (with respect to V) throughout $[0, T]$ and that $g_i(t, \cdot)$, $i \in I$ are strictly invex at $\overline{x}(t)$ (with respect to V) throughout $[0, T]$ with the same $\eta(x(t), \overline{x}(t))$ for all functions. Further, suppose that there exist $\overline{\mu} \in \mathbb{R}^p, \overline{\lambda} \in L_\infty^m[0, T]$ such that*

$$\int_0^T [\sum_{j \in J} \overline{\mu}_j f_j^0(t, \overline{x}(t); h(t)) + \sum_{i \in I} \overline{\lambda}_i(t) g_i^0(t, \overline{x}(t); h(t))]dt \geq 0, \forall h \in L_\infty^n[0, T] \quad (17)$$

$$\overline{\mu}_j \geq 0, j \in J, \overline{\lambda}_i(t) \geq 0 \text{ a.e. in } [0, T], i \in I \quad (18)$$

$$(\overline{\mu}, \overline{\lambda}(t)) \neq (0, 0) \text{ a.e. in } [0, T] \quad (19)$$

$$\overline{\lambda}_i(t) g_i(t, \overline{x}(t)) = 0 \text{ a.e. in } [0, T], i \in I. \quad (20)$$

Then, $\overline{x}$ is a (global) weak efficient solution of (VCNP).

Proof. Suppose to the contrary that $\overline{x}$ is not a weakly efficient solution of (VCNP). Then, there exist $\widetilde{x} \in F$, $\widetilde{x}(t) \neq \overline{x}(t)$ a. e. in $[0, T]$ and such that

$$\int_0^T [f_j(t, \widetilde{x}(t)) - f_j(t, \overline{x}(t))]dt < 0 \quad (21)$$

for all $j \in J$. Since $f_j(t, \cdot)$, $j \in J$ are invex and $g_i(t, \cdot)$, $i \in I$ are strictly invex at $\overline{x}(t)$ throughout $[0, T]$ we have the following inequalities

$$f_j(t, \widetilde{x}(t)) - f_j(t, \overline{x}(t)) \geq f_j^0(t, \overline{x}(t); \eta(\widetilde{x}(t), \overline{x}(t)) \text{ a.e. in } [0, T] \quad (22)$$

$$g_i(t, \widetilde{x}(t)) - g_i(t, \overline{x}(t)) > g_i^0(t, \overline{x}(t); \eta(\widetilde{x}(t), \overline{x}(t)) \text{ a.e. in } [0, T] \quad (23)$$

for all $i \in I, j \in J$. Because $\overline{x} \in \mathbb{F}$ and $\overline{\lambda}_i(t) \geq 0$ a.e. in $[0, T], i \in I$ it is clear that

$$\overline{\lambda}_i(t) g_i(t, \overline{x}(t)) \leq 0 \text{ a.e. in } [0, T], \ i \in I. \quad (24)$$

From (18)-(24) it follows that:

$$0 > \int_0^T [\sum_{j \in J} \overline{\mu}_j f_j^0(t, \overline{x}(t); \eta(\widetilde{x}(t), \overline{x}(t)) + \sum_{i \in I} \overline{\lambda}_i(t) g_i^0(t, \overline{x}(t); \eta(\widetilde{x}(t), \overline{x}(t))]$$

and, because $h = \eta(\widetilde{x}(\cdot), \overline{x}(\cdot))$ is in $L_\infty^n[0, T]$ the last inequality contradicts (17). Therefore, we conclude that $\overline{x}$ is a global weak efficient solution of (VCNP). $\blacksquare$

Remark 12 *From the above proof, it is clear that, if $f_j(t, \cdot), g_i(t, \cdot)$ are invex for each i and $j \in J$ and, at least one $k \in I$, say, $g_k(t, \cdot)$ are strictly invex at $\overline{x}(t)$ throughout $[0, T]$ and such that the corresponding multiplier $\overline{\lambda}_k$ is nonzero on a subset of $[0, T]$ which have positive Lebesgue measure, then the assertion of the Theorem 11 remains valid.*

If the functions of (VCNP) are invex, the Karush-Kuhn-Tucker conditions are sufficient for the weak efficiency. In effect:

Theorem 13 *[Karush-Kuhn-Tucker sufficiency] Let $\overline{x} \in \mathbb{F}$ be a given point. Suppose that $f_j(t, \cdot), g_i(t, \cdot)$ are invex at $\overline{x}(t)$ (with respect to V) throughout $[0, T]$ with the same $\eta(x(t), \overline{x}(t))$ for all functions. Further, suppose that there exist $\overline{\mu} \in \mathbb{R}^p, \overline{\mu} \neq 0$ and $\lambda \in L_\infty^m[0, T]$ such that*

$$\int_0^T [\sum_{j \in J} \overline{\mu}_j f_j^0(t, \overline{x}(t); h(t)) + \sum_{i \in I} \overline{\lambda}_i(t) g_i^0(t, \overline{x}(t); h(t))] dt \geq 0, \forall h \in L_\infty^n[0, T] \quad (25)$$

$$\overline{\mu}_j \geq 0, j \in J, \overline{\lambda}_i(t) \geq 0 \text{ a.e. in } [0, T], i \in I \quad (26)$$

$$\overline{\lambda}_i(t) g_i(t, \overline{x}(t)) = 0 \text{ a.e. in } [0, T], i \in I. \quad (27)$$

then $\overline{x}$ is a (global) weakly efficient solution of (VCNP).

Proof. Let $\overline{x} \in \mathbb{F}$ be given. It follows from (26)-(27) that $\overline{\lambda}_i g_i(t, x(t)) \leq \overline{\lambda}_i(t) g_i(t, \overline{x}(t)) = 0$, for all $x \in F$. Since for each $i \in I$, $g_i(t, \cdot)$ is invex at $\overline{x}(t)$, then $\overline{\lambda}_i(t) g_i(t, \cdot)$ is also invex at $\overline{x}(t)$ throughout $[0, T]$ for the same η. From the invexity of $\overline{\lambda}_i(t) g_i(t, \cdot)$, we obtain

$$\overline{\lambda}_i(t) g_i^0(t, \overline{x}(t); \eta(x(t), \overline{x}(t))) \leq 0 \text{ a.e. in } [0, T], i \in I. \quad (28)$$

Now, setting $h(t) = \eta(x(t), \overline{x}(t))$ in (15) we get

$$0 \leq \int_0^T [\sum_{j \in J} \overline{\mu}_j f_j^0(t, \overline{x}(t); \eta(x(t), \overline{x}(t))) + \sum_{i \in I} \overline{\lambda}_i(t) g_i^0(t, \overline{x}(t); \eta(x(t), \overline{x}(t)))] dt. \quad (29)$$

Combining (28) and (29) we obtain

$$\int_0^T [\sum_{j \in J} \overline{\mu}_j f_j^0(t, \overline{x}(t); \eta(x(t), \overline{x}(t)))] dt \geq 0. \quad (30)$$

Suppose that $\overline{x}$ is not a weakly efficient solution of (VCNP). Then there exist $x \in F, x \neq \overline{x}$ such that

$$\int_0^T [f_j(t, x(t)) - f_j(t, \overline{x}(t))] dt < 0.$$

Hence, $\overline{\mu} \in \mathbb{R}_+^p \backslash \{0\}$ together with the invexity of f_j, we conclude that

$$\int_0^T [\sum_{j \in J} \overline{\mu}_j f_j^0(t, \overline{x}(t); \eta(x(t), \overline{x}(t)))] dt < 0$$

which contradicts (30). Therefore, $\overline{x}$ is a global weakly efficient solution of (VCNP). ■

Next we will prove that if the functions are invex and the multipliers corresponding with the objective functions are positive, then $\overline{x}$ is a properly efficient solution. In effect:

Theorem 14 *Let $\overline{x} \in \mathbb{F}$ be a given point. Suppose that $f_j(t, \cdot), g_i(t, \cdot)$ are invex at $\overline{x}(t)$ (with respect to V) throughout $[0, T]$ with the same $\eta(x(t), \overline{x}(t))$ for all functions. Further, suppose that there exist $\overline{\mu} \in \mathbb{R}^p$ and $\lambda \in L_\infty^m[0, T]$ such that*

$$\int_0^T [\sum_{j \in J} \overline{\mu}_j f_j^0(t, \overline{x}(t); h(t)) + \sum_{i \in I} \overline{\lambda}_i(t) g_i^0(t, \overline{x}(t); h(t))] dt \geq 0, \forall h \in L_\infty^n[0, T] \quad (31)$$

$$\overline{\mu}_j > 0, j \in J, \overline{\lambda}_i(t) \geq 0 \ a.e. \ in \ [0, T], i \in I \quad (32)$$

$$\overline{\lambda}_i(t) g_i(t, \overline{x}(t)) = 0 \ a.e. \ in \ [0, T], i \in I. \quad (33)$$

then $\overline{x}$ is a properly efficient solution of (VCNP).

Proof. First we shall prove that $\overline{x}$ is an efficient solution of (VCNP). In fact, suppose that $\overline{x}$ is not an efficient solution of (VCNP), that is, there exists $x \in F$ such that

$$\int_0^T f_j(t, x(t)) dt \leq \int_0^T f_j(t, \overline{x}(t)) dt, \forall j \in J \quad (34)$$

with a strict inequality valid for some j. Therefore, since $\overline{\mu}_j > 0$, (34) implies that

$$\int_0^T \sum_{j \in J} \overline{\mu}_j [f_j(t, x(t)) - f_j(t, \overline{x}(t))] dt < 0. \quad (35)$$

It is easy to see that

$$0 \geq \overline{\lambda}_i(t) g_i(t, x(t)) = \overline{\lambda}_i(t) g_i(t, x(t)) - \overline{\lambda}_i(t) g_i(t, \overline{x}(t)) \quad (36)$$

for all $i \in I$. The invexity hypothesis on $f_j(t, \cdot)$ and $g_i(t, \cdot)$ together (35) and (36) imply that

$$0 > \int_0^T [\sum_{j \in J} \overline{\mu}_j f_j^0(t, \overline{x}(t); \eta(x(t), \overline{x}(t))) + \sum_{i \in I} \overline{\lambda}_i(t) g_i^0(t, \overline{x}(t); \eta(x(t), \overline{x}(t)))$$

which contradicts (31). Hence, $\overline{x}$ is an efficient solution of (VCNP).

Now, define: $M = (p-1)\max\limits_{i,j\in J}\dfrac{\overline{\mu}_i}{\overline{\mu}_j}$ (note that we are supposing that $p \geq 2$). We will prove that $\overline{x}$ is a properly efficient solution of (VCNP) with respect to this M.

In fact, if $\overline{x}$ is not a properly efficient solution with respect to M, then there exist $i \in I$ and $x \in F$ such that

$$\phi_i(\overline{x}) - \phi_i(x) > M[\phi_j(x) - \phi_j(\overline{x})]$$

for each $j \in J$ satisfying $\phi_j(x) > \phi_j(\overline{x})$. Hence,

$$\phi_i(\overline{x}) - \phi_i(x) > M[\phi_j(x) - \phi_j(\overline{x})], \forall j \neq i.$$

Thereforem

$$\phi_i(\overline{x}) - \phi_i(x) > (p-1)\frac{\overline{\mu}_j}{\overline{\mu}_i}[\phi_j(x) - \phi_j(\overline{x})]$$

that is,

$$\frac{\overline{\mu}_i}{p-1}[\phi_i(\overline{x}) - \phi_i(x)] > \overline{\mu}_j[\phi_j(x) - \phi_j(\overline{x})], \forall j \neq i.$$

Summing on $j \neq i$, we obtain

$$\overline{\mu}_i[\phi_i(\overline{x}) - \phi_i(x)] > \sum_{\substack{j\in J \\ j\neq i}} \overline{\mu}_j[\phi_j(x) - \phi_j(\overline{x})]$$

and hence

$$\sum_{j\in J} \overline{\mu}_j[\phi_j(x) - \phi_j(\overline{x})] < 0.$$

From the inequality above, the invexity conditions and (31)-(33), we conclude that

$$
\begin{aligned}
0 &\leq \int_0^T [\sum_{j\in J} \overline{\mu}_j f_j^0(t, \overline{x}(t); \eta(x(t), \overline{x}(t))) + \sum_{i\in I} \overline{\lambda}_i(t) g_i^0(t, \overline{x}(t); \eta(x(t), \overline{x}(t)))]dt \\
&\leq \int_0^T \{\sum_{j\in J} \overline{\mu}_j[f_j(t, x(t)) - f_j(t, \overline{x}(t))] + \sum_{i\in I} \overline{\lambda}_i(t)[g_i(t, x(t)) - g_i(t, \overline{x}(t))]\}dt \\
&= \sum_{j\in J} \overline{\mu}_j(\phi_j(x) - \phi_j(\overline{x})) + \int_0^T \sum_{i\in I} \overline{\lambda}_i(t) g_i(t, x(t))dt \leq \sum_{j\in J} \overline{\mu}_j(\phi_j(x) - \phi_j(\overline{x})) < 0
\end{aligned}
$$

which is a contradiction. Therefore, $\overline{x}$ is a properly efficient solution of (VCNP). ∎

In the sequel $L'_x(x, \mu, \lambda; h)$ denotes the usual derivative of the Lagrangean function $L(\cdot, \mu, \lambda)$ at x in the direction $h \in L_\infty^n[0, T]$ and $\partial_x L(x, \mu, \lambda)$ means the generalized gradient of $L(\cdot, \mu, \lambda)$.

We point out the conditions (17)-(20) in Theorem 11 (or (25)-(27) in Theorem 13) cannot be written in terms of the Clarke generalized of the Lagrangean function, in general. In the follow, we show that under regularity assumptions, it is possible. In fact, if f_j and g_i, $i \in I, j \in J$ are Clarke regular at $\overline{x}$, then condition (17) is equivalent to $L'_x(\overline{x}, \overline{\mu}, \overline{\lambda}; h) \geq 0$, $\forall h \in L_\infty^n[0, T]$ and therefore, $0 \in \partial_x L(\overline{x}, \overline{\mu}, \overline{\lambda})$. Formally, we have the following corollaries:

Corollary 15 *Let $\overline{x} \in \mathbb{F}$ and suppose that $f_j(t, \cdot)$, $j \in J$, $g_i(t, \cdot)$, $i \in I$ are Clarke regular. Further, suppose that $f_j(t, \cdot)$ are invex at $\overline{x}(t)$ and $g_i(t, \cdot)$ are strictly invex at $\overline{x}(t)$ (with respect to V) throughout $[0, T]$ with the same $\eta(x(t), \overline{x}(t))$, for all $i \in I, j \in J$ and suppose that there exist $\overline{\mu} \in \mathbb{R}^p, \overline{\lambda} \in L^m_\infty[0, T]$ such that*

$$0 \in \partial_x L(\overline{x}, \overline{\mu}, \overline{\lambda})$$
$$\overline{\mu}_j \geq 0, j \in J, \overline{\lambda}_i(t) \geq 0 \ a.e. \ in \ [0, T], \ i \in I$$
$$(\overline{\mu}, \overline{\lambda}(t)) \neq (0, 0) \ a.e. \ in \ [0, T]$$
$$\overline{\lambda}_i(t)g_i(t, \overline{x}(t)) = 0 \ a.e. \ in \ [0, T], i \in I$$

then $\overline{x}$ is a (global) weak efficient solution of (VCNP).

Corollary 16 *Let $\overline{x} \in \mathbb{F}$ and suppose that $f_j(t, \cdot)$, $j \in J$, $g_i(t, \cdot)$, $i \in I$ are Clarke regular and invex at $\overline{x}(t)$ (with respect to V) throughout $[0, T]$ with the same $\eta(x(t), \overline{x}(t))$ for all functions. Further, suppose that there exist $\overline{\mu} \in \mathbb{R}^p, \overline{\mu} \neq 0, \overline{\lambda} \in L^m_\infty[0, T]$ such that*

$$0 \in \partial_x L(\overline{x}, \overline{\mu}, \overline{\lambda})$$
$$\overline{\mu}_j \geq 0, j \in J, \overline{\lambda}_i(t) \geq 0 \ a.e. \ in \ [0, T], \ i \in I$$
$$\overline{\lambda}_i(t)g_i(t, \overline{x}(t)) = 0 \ a.e. \ in \ [0, T], i \in I$$

then $\overline{x}$ is a (global) weak efficient solution of (VCNP).

Corollary 17 *Let $\overline{x} \in \mathbb{F}$ and suppose that $f_j(t, \cdot)$, $j \in J$, $g_i(t, \cdot)$, $i \in I$ are Clarke regular and invex at $\overline{x}(t)$ (with respect to V) throughout $[0, T]$ with the same $\eta(x(t), \overline{x}(t))$ for all functions. Further, suppose that there exist $\overline{\mu} \in \mathbb{R}^p, \overline{\mu} \neq 0, \overline{\lambda} \in L^m_\infty[0, T]$ such that*

$$0 \in \partial_x L(\overline{x}, \overline{\mu}, \overline{\lambda})$$
$$\overline{\mu}_j > 0, j \in J, \overline{\lambda}_i(t) \geq 0 \ a.e. \ in \ [0, T], \ i \in I$$
$$\overline{\lambda}_i(t)g_i(t, \overline{x}(t)) = 0 \ a.e. \ in \ [0, T], i \in I$$

then $\overline{x}$ is a properly efficient efficient solution of (VCNP).

5   The differentiable case

In scalar optimization theory, the Kuhn-Tucker conditions are sufficient for optimality if the functions involved are invex. For constrained problems, the invexity defined by Hanson [7] is a sufficient condition but not a necessary condition for every Kuhn-Tucker stationary point to be a global minimizer. Martin [9] defined a weaker invexity notion, called Kuhn-Tucker invexity of KT-invexity, which is both necessary and sufficient to establish the Kuhn-Tucker optimality conditions in scalar programming problems. Recently, Osuna Gómez *et al.* [11] proposed a notion of KT-invexity for smooth multiobjective problems and obtained results similar with those obtained by Martin. In this Section, we propose the concept of KT-pseudoinvexity and we extend the results obtained by Martin [9] and Osuna-Gómez *et al.* [11] for multiobjective continuous-time problems.

In this Section, will suppose that the functions $t \mapsto f_j(x(t), t)$, $j \in J$, and $t \mapsto g_i(x(t), t)$, $i \in I$, are Lebesgue measurable and integrable for all $x \in X$. We assume also that the functions $f_j, j \in J$, and g_i, $i \in I$, are continuously differentiable with respect to their first arguments.

In what follows we state a result which will be useful for the proof of our results. This result can be viewed as a Generalized Motzkin Theorem of the alternative. It is the continuous-time analogue of the theorem given on page 66 of the book by Mangasarian [8] and its proof is almost identical to the one given in the Mangasarian's book.

Theorem 18 *Let $Z \subseteq L_\infty^n[0, T]$ be a nonempty convex subset. Let $p : W \times [0, T] \to \mathbb{R}^m$ and $q : W \times [0, T] \to \mathbb{R}^k$ be mappings given by $p(z(t), t) = \pi(z)(t)$ and $q(z(t), t) = B(t)z(t) - b(t)$, respectively, where $W \subseteq \mathbb{R}^n$ is an open subset, π is a mapping from Z into $\Lambda_1^m[0, T]$, $B(t)$ is a $k \times n$ matrix and $b(t) \in \mathbb{R}^k$. We assume that p is convex with respect to its first argument in W throughout $[0, T]$ and that there does not exist $v \in L_\infty^k[0, T] \setminus \{0\}$, $v(t) \geq 0$ a.e. in $[0, T]$, such that*

Theorem 19

$$B'(t)v(t) = 0 \ \text{ a.e. in } [0, T]. \tag{37}$$

Then exactly one of the following systems is consistent:

(I) $p(z(t), t) < 0$, $B(t)z(t) \leq b(t)$ a.e. in $[0, T]$ has a solution $z \in Z$;

(II) $\displaystyle\int_0^T \{u'(t)p(z(t), t) + v'(t)[B(t)z(t) - b(t)]\}dt \geq 0$ for all $z \in Z$, for some $u \in L_\infty^m[0, T]$, $u(t) \geq 0$, $u(t) \neq 0$ a.e. in $[0, T]$ and for some $v \in L_\infty^k[0, T]$, $v(t) \geq 0$ a.e. in $[0, T]$.

Proof. Quite similar to the proof of the Theorem 3.4, page 137 in [17]. ∎

Definition 20 *A feasible solution y is said to be a vector Karush-Kuhn-Tucker solution (or vector KKT-solution) for Problem (CMP) if there exists $\mu \in \mathbb{R}^p$ and $\lambda \in L_\infty^m[0, T]$ such that*

$$\int_0^T \left[\sum_{j \in J} \mu_j \nabla f_j'(y(t), t) + \sum_{i \in I} \lambda_i(t) \nabla g_i'(y(t), t) \right] h(t)dt = 0 \ \forall \ h \in L_\infty^n[0, T], \tag{38}$$

$$\lambda_i(t)g_i(y(t), t) = 0 \text{ a.e. in } [0, T], \ i \in I, \tag{39}$$

$$\lambda_i(t) \geq 0 \text{ a.e. in } [0, T], \ i \in I, \tag{40}$$

$$\mu_j \geq 0, \ j \in J, \text{ and } \mu \neq 0. \tag{41}$$

At last we give a constraint qualification in the continuous-time setting.

Definition 21 *We say that the constraints* g_i, $i \in I$, *satisfy (CQ) at* $y \in \mathbb{F}$ *if there do not exist* $v_i \in L_\infty[0, T]$, $v_i(t) \geq 0$ *a.e. in* $[0, T]$, $i \in I$, *not all zero, such that*

$$\sum_{i \in I} \int_{A_i(y)} v_i(t) \nabla g_i(y(t), t) h(t) dt \geq 0 \ for \ all \ h \in L_\infty^n[0, T].$$

In this section we introduce the notion of Karush-Kuhn-Tucker pseudoinvexity for (CMP). Further, we state and prove a result which provides necessary and sufficient conditions for global optimality of a vector Karush-Kuhn-Tucker solution.

Definition 22 *The Problem (CMP) is said to be Karush-Kuhn-Tucker pseudoinvex (or KKT-pseudoinvex) if there exists a function* $\eta : V \times V \times [0, T] \to \mathbb{R}^n$ *such that* $t \mapsto \eta(x(t), y(t), t) \in L_\infty^n[0, T]$ *and*

$$\phi(x) < \phi(y) \Rightarrow \int_0^T \nabla f_j'(y(t), t) \eta(x(t), y(t), t) dt < 0, \ j \in J, \tag{42}$$

$$-\nabla g_i'(y(t), t) \eta(x(t), y(t), t) \geq 0 \ a.e. \ in \ A_i(y), \ i \in I, \tag{43}$$

for all $x, y \in \mathbb{F}$.

Theorem 23 *We assume that the constraints* g_i, $i \in I$, *satisfy (CQ) at each* $y \in \mathbb{F}$. *Then every vector KKT-solution is a weak efficient solution of (CMP) if and only if (CMP) is KKT-pseudoinvex.*

Proof. Let y be a vector KKT-solution and suppose that (CMP) is KKT-pseudoinvex. Suppose that there exists a feasible solution x such that $\phi(x) < \phi(y)$. As (CMP) is KKT-pseudoinvex, using (42), we obtain

$$\int_0^T \nabla f_j'(y(t), t) \eta(x(t), y(t), t) dt < 0, \ j \in J. \tag{44}$$

Since y is a vector KKT-solution there exist $\mu \in \mathbb{R}^p$ and $\lambda \in L_\infty^m[0, T]$ satisfying (38)-(41). By (41) and (44) we have

$$\int_0^T \sum_{j \in J} \mu_j \nabla f_j'(y(t), t) \eta(x(t), y(t), t) dt < 0.$$

Using (38) with $h(t) = \eta(x(t), y(t), t)$, $t \in [0, T]$, we obtain

$$\int_0^T \sum_{i \in I} \lambda_i(t) \nabla g_i'(y(t), t) \eta(x(t), y(t), t) dt > 0. \tag{45}$$

By the other hand, from (40) and (43), since by (39) $\lambda_i(t) = 0$, $t \notin A_i(y)$, $i \in I$, it follows that

$$\int_0^T \sum_{i \in I} \lambda_i(t) \nabla g_i'(y(t), t) \eta(x(t), y(t), t) dt \leq 0,$$

which is a contradiction to (45). Therefore y is a weakly efficient solution.

Conversely, suppose that every vector KKT-solution is a weakly efficient solution. Let $x, y \in \mathbb{F}$ be such that $\phi(x) < \phi(y)$. Then y is not a weakly efficient solution, so that, by hypothesis, y is not a vector Karush-Kuhn-Tucker solution. So the system

$$\int_0^T \left[\sum_{j \in J} \mu_j \nabla f_j'(y(t), t) + \sum_{i \in I} \lambda_i(t) \nabla g_i'(y(t), t) \right] h(t) dt = 0 \ \forall \ h \in L_\infty^n[0, T],$$

$$\lambda_i(t) g_i(y(t), t) = 0 \text{ a.e. in } [0, T], \ i \in I,$$

$$\lambda_i(t) \geq 0 \text{ a.e. in } [0, T], \ i \in I,$$

$$\mu_j \geq 0, \ j \in J, \text{ and } \mu \neq 0,$$

has no solution $(\mu, \lambda) \in \mathbb{R}^p \times L_\infty^m[0, T]$. Equivalently, the system

$$\int_0^T \left[\sum_{j \in J} \mu_j \nabla f_j'(y(t), t) + \sum_{i \in I} \lambda_i(t) \chi_i(t) \nabla g_i'(y(t), t) \right] h(t) dt = 0 \ \forall \ h \in L_\infty^n[0, T]$$

$$\lambda_i(t) \geq 0 \text{ a.e. in } [0, T], \ i \in I,$$

$$\mu_j \geq 0, \ j \in J, \text{ and } \mu \neq 0,$$

has no solution $(\mu, \lambda) \in \mathbb{R}^p \times L_\infty^m[0, T]$, where $\chi_i : [0, T] \to \mathbb{R}$ is defined, for each $i \in I$, by

$$\chi_i(t) = \begin{cases} 1 & \text{if } t \in A_i(y), \\ 0 & \text{if } t \notin A_i(y). \end{cases}$$

As the constraint qualification holds by hypothesis, the condition (37) in Theorem 18 is verified. Applying that theorem, it follows that there exists $h \in L_\infty^n[0, T]$ such that

$$\int_0^T \nabla f_j'(y(t), t) h(t) dt < 0, \ j \in J,$$

$$\chi_i(t) \nabla g_i'(y(t), t) h(t) \leq 0 \text{ a.e. in } [0, T], \ i \in I.$$

Define $\eta(x(t), y(t), t) = h(t)$ a.e. in $[0, T]$. Therefore

$$\int_0^T \nabla f_j'(y(t), t) \eta(x(t), y(t), t) dt < 0, \ j \in J,$$

$$-\nabla g_i'(y(t), t) \eta(x(t), y(t), t) \geq 0 \text{ a.e. in } A_i(y), \ i \in I.$$

Thus there exists a function $\eta : V \times V \times [0, T] \to \mathbb{R}^n$ such that $t \mapsto \eta(x(t), y(t), t) \in L_\infty^n[0, T]$ and

$$\phi(x) < \phi(y) \Rightarrow \int_0^T \nabla f_j'(y(t), t) \eta(x(t), y(t), t) dt < 0, \ j \in J,$$

$$-\nabla g_i'(y(t), t) \eta(x(t), y(t), t) \geq 0 \text{ a.e. in } A_i(y), \ i \in I,$$

for all $x, y \in \mathbb{F}$, so that (CMP) is KKT-pseudoinvex. ∎

References

[1] J. Aubin and H. Frankowska, Set Valued Analysis, Birkhauser, Boston, 1990.

[2] R. Bellman, Bottleneck problems and dynamic programming. Proc. Nat. Acad. Sci. U. S. A., 39 (1953) 947-951.

[3] A.J.V. Brandão, M. Rojas-Medar and G.N. Silva, Nonsmooth continuous-time optimization problems: necessary conditions, Comput. Math. Appl. 41 (2001) 1477-1486.

[4] F.H. Clarke, Optimization and Nonsmooth Analysis, Wiley, New York, 1983.

[5] B.D. Craven, Nondifferentiable optimization by smooth approximations, Optimization 17 (1986) 3-17.

[6] A.M. Geoffrion, Proper efficiency and the theory of vector maximization, J. Math. Anal. Appl. 80 (1981) 545-550.

[7] M.A. Hanson, On sufficiency of the Kuhn-Tucker conditions, J. Math. Anal. Appl. 80 (1981) 545-50.

[8] O.L. Mangasarian, Nonlinear programming, Classics in Applied Mathematics, SIAM, Philadelphia USA, 1994.

[9] D.H. Martin, The Essence of Invexity, J. Optim. Theory Appl. 47 (1985) 65-76.

[10] V.A. Oliveira and M.A. Rojas-Medar, Continuous-time multiobjective optimization problems via invexity, Abstract and Applied Analysis, Volume 2007, article ID 61296, 11 pages. doi:101155/2007/61296.

[11] R. Osuna-Gómez, A. Rufián-Lizana and P. Ruíz-Canales, Invex functions and generalized convexity in multiobjective programming, J. Optim. Theory Appl. 98 (3)(1998) 651-661.

[12] T.W. Reiland. Nonsmooth invexity, Bull. Austral. Math. Soc. 42 (1990) 437-446.

[13] M.A. Rojas-Medar, A.J.V. Brandão and G. N. Silva, Nonsmooth continuous-time optimization problems: sufficient conditions, J. Math. Anal. Appl. 227 (1998) 305-318.

[14] L.B. Santos, A.J.V. Brandão, R. Osuna-Gómez and M.A. Rojas-Medar, Nonsmooth continuous-time multiobjective optimization problems, Tech. Rep., State University of Campinas, São Paulo, Brazil, 2005.

[15] G.J. Zalmai, The Fritz-John and Kuhn-Tucker optimality conditions in continuous-time nonlinear programming, J. Math. Anal. Appl. 110 (1985) 503-518.

[16] G.J. Zalmai, Optimality conditions and Lagrangean duality in continuous-time nonlinear programming, J. Math. Anal. Appl. 109 (1985) 426-452.

[17] G.J. Zalmai, Continuous-time generalization of Gordan´s transposition theorem, J. Math. Anal. Appl. 110 (1985) 130-140.

CHAPTER 10

Optimality conditions and duality for nonsmooth multiobjective continuous-time problems

S. Nobakhtian *   M.R. Pouryayevali,[†]

Abstract

In this chapter we present some classes of nonsmooth continuous-time problems. Optimality conditions under certain structure of generalized convexity are derived for these classes. Subsequently, two dual models are formulated and weak and strong duality theorems are established.

Keywords: Continuous-time problems, duality.

1   Introduction

Continuous-time problems, were introduced initially by Bellman [3]. He considered a type of optimization problem, which is now known as continuous-time linear programming, formulated its dual and provided duality relations. Two types of problems fitting into this scheme are variational and control problems.

On the other hand, investigation of optimality conditions and/or duality has been one of the most attracting topics in the theory of nonlinear programming. Several classes of optimization problems with multiobjective functions have been the subject of numerous investigations in the past few years. Problems of multiobjective optimization are widespread in mathematical modelling of real word systems for a very broad range of applications. For instance, multiobjective optimization problems which arise in mechanical engineering are discussed in Stadler [26]. Applications of multiobjective optimization techniques for the design of aircraft control systems are given in Schy and Giesy [25]. Various other applications of multiobjective optimization in resource planning and management, mathematical biology, and in welfare economic can be found in Stadler [27].

As a consequence, this intense interest has led to the development of various optimality conditions and duality results for *multiobjective continuous-time* problem; see e.g. [28, 29]. In [29] Zalmai established parametric and semi parametric stationary point-type and saddle-point-type necessary and sufficient proper efficiency conditions for the

---
*Department of Mathematics, University of Isfahan, P.O. Box 81745-163, Isfahan, Iran.   e-mail: nobakht@math.ui.ac.ir

[†]Department of Mathematics, University of Isfahan, P.O. Box 81745-163, Isfahan, Iran.   e-mail: pourya@math.ui.ac.ir

continuous-time multiobjective fractional programming problems with nonsmooth convex data (nondifferentiable convex functions). However, from a survey of publications in this area, it appears that nonsmooth multiobjective continuous-time problems have not received much attention in the related literature.

Convexity plays an important role in deriving sufficient optimality conditions and duality results. Convexity assumptions are often not satisfied in real-word problems as for instance in *economic models*, see [2]. In the last few years, attempts have been made to weaken the convexity hypotheses and thus explore the extent of optimality conditions applicability. The *invexity* idea was introduced by Hanson [13] and Craven [9]. Hanson [13] defined invex functions allowing the use of the *Kuhn-Tucker* conditions as sufficient conditions for optimality in constrained optimization problems. Since the invexity condition for global optimality is substantially weaker than convexity, it may be expected to hold for a general class of problems, for instance in economic models; see [17].

In this Chapter, we introduce the concept of invexity to the continuous time context and use it to obtain optimality conditions and duality results for nonsmooth multiobjective continuous-time problems. For this purpose the chapter is divided in four sections. In Section 2, we introduce the concept of invexity to continuous functions. In Section 3, we have considered a class of nonsmooth multiobjective continuous-time problems and establish a number of Kuhn-Tucker type optimality conditions and discuss an example. In Section 4, we formulate two duality type problems and prove appropriate duality theorems. In our approach, the usual convexity requirement and differentiability for the functions are relaxed.

Throughout this chapter we consider the following multiobjective continuous-time problem(MCTP):

$$\min \quad \phi(x) = \int_0^T f(t, x(t))dt = (\int_0^T f_1(t, x(t))dt, ..., \int_0^T f_r(t, x(t))dt)$$

$$s.t. \quad g_j(t, x(t)) \leqq 0, j \in M = \{1, 2, ..., m\}, \ x \in X, \ a.e. \ t \in [0, T].$$

Here X is a nonempty open convex subset of the Banach space $(C[0, T], \mathbb{R}^n)$ of all $n-$dimensional vector-valued continuous functions defined on the compact interval $[0, T] \subset \mathbb{R}$, with the norm $\| \cdot \|_\infty$ defined by

$$\|x\|_\infty = \max_{1 \leq j \leq n} \text{ess} \ \sup\{|x_j(t)|, 0 \leq t \leq T\},$$

where for each $t \in [0, T], x_j(t)$ is the jth component of $x(t) \in \mathbb{R}^n$. The function $\phi(.)$ is a vector valued function defined on $X, , g_j(t, x(t)) = G_j(x)(t), j \in M$ and $f_i(t, x(t)) = F_i(x)(t), i \in L = \{1, 2, ..., r\}$, where G_i and F_i are maps form X into the Banach space $(C[0, T], \mathbb{R}^n)$. Let F_p be the set of feasible solutions of (MCTP),

$$F_P = \{x \in X : g_j(t, x(t)) \leq 0, a.e \ \ t \in [0, T], j \in M\}.$$

We assume that functions $t \to f_i(t, x(t))$, and $t \to g_j(t, x(t))$, are Lebesgue measurable for all $x \in X$. We also assume that for every (fixed) $t \in [0, T]$ the functions $f_i(t, x(t)$ and $g_j(t, x(t))$ are locally Lipschitz on X.

Throughout this chapter, $\Lambda_1^m[0,T]$ denote the space of all Lebesgue measurable essentially bounded $m-$dimensional vector functions defined on $[0,T]$, with the norm $\|\cdot\|_1$ defined by

$$\|y\|_1 = \max_{1\leq i\leq m} \int_0^T |y_i(t)|dt.$$

We recall basic concepts and tools from nonsmooth analysis. Most of the material included here can be found in [6, 7].

Let Y be a Banach space and $F:Y\to R$ be a locally Lipschitz function. The generalized gradient of F at x, denoted by $\partial_c F(x)$, is defined by

$$\partial_c F(x) := \{\xi \in Y^* : \langle \xi, d\rangle \leq F^0(x;d) \quad \forall \quad d \in Y\}.$$

Here Y^* denotes the dual space of continuous linear functionals on Y, and $\langle \cdot, \cdot\rangle : Y^* \times Y \to \mathbb{R}$ is the duality pairing.

Now, assume $\bar{x} \in X$ and $h \in L_\infty^n[0,T]$, where $L_\infty^n[0,T]$ is the space of all $n-$dimensional vector-valued Lebesgue measurable essentially bounded functions defined on the compact interval $[0,T] \subset \mathbb{R}$, with the norm $\|\cdot\|_\infty$.

Brandao et al. [4, 5] defined the continuous Clarke generalized directional derivative of f_i as follows;

$$f_i^0(t,\bar{x}(t);h(t)) := F_i^0(\bar{x};h)(t) := \limsup_{\substack{y \to \bar{x} \\ \lambda \downarrow 0}} \frac{F_i(y+\lambda h)(t) - F_i(y)(t)}{\lambda},$$

a.e. $t \in [0,T]$.

Let us recall the integration of multifunctions. Given a multifunction $H:[0,T]\to\mathbb{R}^n$, denote by S(H) the following set:

$$S(H) = \{f \in L_1^n[0,T], f(t) \in H(t) \quad a.e. \quad t \in [0,T]\}.$$

We define the integral of H, denoted by $\int_0^T H(t)dt$, as the following subset of $\mathbb{R}^n$:

$$\int_0^T H(t)dt := \left\{ \int_0^T f(t)dt : f \in S(H)\right\}.$$

2   Invexity for continuous-time problems

We introduce invexity notions in the continuous-time context. Let f be a real function on $[0,T]\times X$ and suppose that $f(t,.)$ is locally Lipschitz on X throughout $[0,T]$. Assume that there exists a function $\eta : X \times X \to \mathbb{R}$ such that the function $t \to \eta(x(t),\bar{x}(t))$ is in $L_\infty^n[0,T]$.

Definition 1 *The function $f(t,.)$ is said to be:*

(i) *invex at $\bar{x}$, with respect to η, if for all $x \in X$,*

$$f(t, x(t)) - f(t, \bar{x}(t)) \geq f^0(t, \bar{x}(t); \eta(x(t), \bar{x}(t)), \quad a.e.\ t \in [0, T],$$

(ii) *strictly invex if the above inequality is strict for $x(t) \neq \bar{x}(t)$ a.e. in $[0, T]$,*

(iii) *pseudoinvex at $\bar{x}$ with respect to η, if for all $x \in X$,*

$$f^0(t, \bar{x}(t); \eta(x(t), x\bar{(}t)) \geq 0, \Rightarrow f(t, x(t)) \geq f(t, \bar{x}(t)), \quad a.e.\ t \in [0, T],$$

(iv) *strictly pseudoinvex at $\bar{x}$ with respect to η, if for all $x(t) \neq \bar{x}(t)$ a.e. in $[0, T]$,*

$$f^0(t, \bar{x}(t); \eta(x(t), x\bar{(}t)) \geq 0 \Rightarrow f(t, x(t)) > f(t, \bar{x}(t)),$$

(v) *quasiinvex at $\bar{x}$ with respect to η, if for all $x \in X$, all $x \in X$,*

$$f(t, x(t)) \leq f(t, \bar{x}(t)) \Rightarrow f^0(t, \bar{x}(t); \eta(x(t), x\bar{(}t)) \leq 0, \quad a.e.\ t \in [0, T],$$

(vi) *strictly quasiinvex at $\bar{x}$ with respect to η, if for all $x \in X$,*

$$f(t, x(t)) \leq f(t, \bar{x}(t)) \Rightarrow f^0(t, \bar{x}(t); \eta(x(t), \bar{x}(t)) < 0, \quad a.e.\ t \in [0, T].$$

We use the acronyms PIX, SPIX, QIX, SQIX for $f(t,.)$ when it is pseudoinvex, strictly pseudoinvex, quasiinvex, and strictly quasiinvex, respectively, at each point of X.

Examples 1 *Consider the function $f(t, x(t)) = |x(t)|$, where $x : [1, 2] \to \mathbb{R}$ is defined by $x(t) = tx, x \in \mathbb{R}$. It is easy to verify that $f(t,.)$ is invex at $\bar{x}(t) \equiv 0$ with respect to η, where η is defined as:*

$$\eta(x(t), \bar{x}(t)) = x(t) - \bar{x}(t).$$

3   Necessary and sufficient optimality conditions

In this section, we present necessary and sufficient criteria of the Kuhn-Tucker type for the problem (MCP). First we summarize briefly the known results of Kuhn-Tucker type optimality conditions for differentiable continuous-time programming problems.

Optimality conditions of the Kuhn-Tucker type were first considered in continuous-time programming by Hanson and Mond [14] for the following linearly constrained nonlinear program:

$$(P) \quad \max \quad \int_0^T f(x(t))dt$$

$$\text{s.t.} \quad B(t)x(t) \leq c(t) + \int_0^t K(t,s)x(s)ds, \quad 0 \leq t \leq T$$

$$x(t) \geq 0,$$

where $x(t) \in \mathbb{R}^n$ is bounded and measurable on $[0,T]$, $B(t)$ is an $m \times n$ matrix piecewise continuous on $[0,T]$, $c(t) \in \mathbb{R}^n$ is piecewise continuous on $[0,T]$, $K(s,t)$ is an $m \times n$ matrix piecewise continuous on $[0,T] \times [0,T]$, and f is a given concave scalar function twice continuously differentiable.

Imposing certain positivity conditions on $B(t), c(t)$, and $K(s,t)$, they showed that for a function $\bar{x}$ to be an optimal solution of the above problem it is necessary and sufficient that there exist an m-vector $\bar{u}(t) \geq 0$ such that

$$B'(t)\bar{u}(t) - \nabla f((\bar{x}(t)) - \int_t^T K'(s,t)\bar{u}(s)ds \geq 0,$$

$$\int_0^T \bar{x}'(t)[B'(t)\bar{u}(t) - \nabla f((\bar{x}(t)) - \int_t^T K'(s,t)\bar{u}(s)ds]dt = 0,$$

$$\int_0^T \bar{u}'(t)[B(t)\bar{x}(t) - c(t) - \int_0^t K(t,s)\bar{x}(s)ds]dt = 0,$$

where prime denotes transposition.

Their method for deriving these optimality conditions was indirect in the sense that it consisted of linearizing the objective function, applying an extended version of Levinson's linear duality result [15] to establish a duality theorem for the nonlinear problem under consideration, and then deducing the Kuhn-Tucker conditions as a consequence of this nonlinear duality theorem.

Farr and Hanson [11] introduced nonlinearity into the constraints and considered the following general form of the above problem:

$$\max \quad \int_0^T f(x(t))dt$$

$$\text{s.t.} \quad g(x(t)) \leq c(t) + \int_0^t K(t,s)h(x(s))ds, \quad 0 \leq t \leq T$$

$$x(t) \geq 0, \quad 0 \leq t \leq T,$$

where $x(t) \in \mathbb{R}^n$ is bounded and measurable on $[0,T]$, f is a concave twice continuously differentiable scalar function, $K(t,s)$ is an $m \times n$ matrix having nonnegative entries with $K(t,s) = 0$ for $s > t, c(t) \geq 0$, and each component of the vector functions $-g$ and h is concave and differentiable.

Assuming some positivity conditions similar to those of Levinson [15], employing a linearization scheme, and invoking a linear duality result due to Grinold [12], they obtained duality for a linearized form of the problem and then with additional conditions on f, established a duality theorem for the nonlinear problem which in turn was used to prove that for a function $\bar{x}$ to be an optimal solution of the above continuous-time nonlinear programming problem, it is necessary and sufficient that there exist an m vector $\bar{u}(t) \geq 0$ such that

$$\nabla g'(\bar{x}(t)\bar{u}(t) - \nabla f((\bar{x}(t)) - \int_t^T \nabla h'(\bar{x}(t)K'(s,t)\bar{u}(s)ds \geq 0,$$

$$\int_0^T \bar{x}'(t)[\nabla g'(\bar{x}(t)\bar{u}(t) - \nabla f((\bar{x}(t)) - \int_t^T \nabla h'(\bar{x}(t)K'(s,t)\bar{u}(s)ds]dt = 0,$$

$$\int_0^T \bar{u}'(t)[g(\bar{x}(t) - c(t) - \int_0^t K(t,s)h(\bar{x}(s))ds]dt = 0.$$

Similar optimality conditions were established by the same authors in [10] for continuous-time programs with nonlinear time-delayed constraints. The results of Farr and Hanson [11] were subsequently improved and generalized by Reiland and Hanson [23].

Later, optimality conditions for continuous-time nonlinear programming problems have been obtained by direct methods. In [1] a certain regularity assumption is used to establish the Kuhn-Tucker conditions for a class of convex programming problems. Reiland [24], employing a continuous-time version of Zangwills constraint qualification [34] introduced in [23], and an infinite-dimensional form of Farkas theorem, established optimality conditions and duality relations for differentiable continuous-time programs.

Zalmai in [32] considered a class of differentiable continuous-time problem of the form:

$$(CT) \quad \min \quad \phi(x) = \int_0^T f(t, x(t)$$
$$g_i(t, x(t)) \leq 0, i \in M = \{1, 2, ..., m\}$$
$$\text{a.e. } t \in [0, T] \quad x \in X,$$

where X is a nonempty open convex subset of the Banach space $L_\infty^n[0, T]$, and developed the continuous-time analogues of the Fritz John and Kuhn-Tucker optimality conditions in the spirit of finite-dimensional nonlinear programming.

The main auxiliary result making this approach possible is a continuous-time version of Gordan's transposition theorem [30]. Let us state the Gordan's theorem.

Let V be an open set in $\mathbb{R}^n$ containing the set $\{x(t) \in \mathbb{R}^n, x \in X, t \in [0, T]$ and F denote the set of all feasible solutions of problem (CT), that is, let

$$F = x \in X : g_i(t, x(t)) \leq 0 \quad i \in M = \{1, 2, ...m\}, \text{ a.e. in } [0, T]$$

For any $\bar{x} \in F$, let $I(\bar{x})$ denote the index set of all the binding constraints at $\bar{x}$, that is, let

$$I(\bar{x}) = \{i \in \{1, 2, ..., m\} : g_i(\bar{x}(t), t) = 0, \quad \text{a.e. in } [0, T]\}.$$

Theorem 1 *Let X be a nonempty convex subset of $L_\infty^n[0,T]$, let the map $p: V \times [0,T] \to \mathbb{R}^m$ be defined by $p(x(t),t) = \Gamma(x)(t)$, where Γ is a map from X into $\Lambda_1^m[0,T]$. Suppose that p is convex with respect to its first argument on V throughout $[0,T]$. Then either*

$$p(x(t),t) < 0, \quad a.e. \quad in \ [0,T]$$

has a solution $x \in X$, or

$$\int_0^T \mu(t)p(x(t),t)dt \geq 0,$$

for all $x \in X$, and for some $\mu \in \Lambda_\infty^m[0,T] \setminus \{0\}, \mu(t) \geq 0, \quad a.e. \quad in \quad [0,T]$; but never both.

Theorem 2 *(Fritz John necessary optimality theorem) Let $\bar{x}$ be a feasible solution of (CT) . Suppose that f and, for each $i \in I(\bar{x})$, g_i are continuously differentiable in their first arguments at $\bar{x}(t)$ throughout $[0,T]$, that for each $i \in \{1,2,...,m\} \setminus I(\bar{x}), g_i$ is continuous in its first argument at $x(t)$ throughout $[0,T]$, and that the functions $t \mapsto \nabla f((\bar{x}),l)$ and $t \mapsto \nabla g((\bar{x}),t)h(t), i \in I(\bar{x})$, are Lebesgue integrable on [0, T] for all $h \in L_\infty^n[0,T]$. If $\bar{x}$ is a local optimal solution of Problem (CT), then there exist $\mu_0 \in \mathbb{R}, \mu_i \in L_\infty[0,T], i \in I(\bar{x})$ such that*

$$\int_0^T [\mu_0 \nabla f((\bar{x}),t) + \sum_{i \in I(\bar{x})} \mu_i(t)\nabla g_i((\bar{x}),t)]h(t)dt = 0$$

$$\forall h \in L_\infty^n[0,T],$$

$$\mu_0 \geq 0, \mu_i(t) \geq 0, \qquad a.e. \quad in \quad [0,T], \ i \in I(\bar{x}), \ (\mu_0, \mu_{I(\bar{x})}(t)) \neq 0.$$

However, by imposing a regularity condition on the constraint functions, the $\mu_0 \in \mathbb{R}$ may without loss of generality be taken as $\mu_0 = 1$ and Kuhn-Tucker condition for (CT) can be obtained. To prove Kuhn-Tucker type necessary condition, the following *Slater's constraint qualification* is needed.

Let X be a convex set in $L_\infty^n[0,T]$, and let g be convex in its first argument on X throughout $[0,T]$. Then g is said to satisfy Slater's constraint qualification (on X) if there exists an $x \in X$ such that $g(x(t),t) < 0$ a.e. in $[0,T]$.

Theorem 3 *(Kuhn-Tucker necessary optimality theorem). Let $X \in F$ and assume that in addition to the hypotheses of the first part of Theorem 2, for each $i \in I(\bar{x}), g_i$ is convex in its first argument on V throughout $[0,T]$ and satisfies Slater's constraint qualification on X. If $\bar{x}$ is a local optimal solution of Problem (CT), then there exist $\mu_i \in L_\infty^n[0,T], i \in I(\bar{x})$, such that*

$$\int_0^T [\nabla f((\bar{x}),t) + \sum_{i \in I(\bar{x})} \mu_i(t)\nabla g_i((\bar{x}),t)]h(t)dt = 0$$

$$\forall h \in L_\infty^n[0,T],$$

$$\mu_i(t) \geq 0, \qquad a.e. \quad in \quad [0,T], \ i \in I(\bar{x}).$$

Since then, various generalization of the Kuhn-Tucker optimality conditions have been established. However, the major difficulty are that the problems require differentiability and convexity assumptions. As nonsmooth phenomena in optimization occur naturally and frequently, the attempts to weaken these smoothness requirements have received a great deal of attention during the last two decades. Necessary and sufficient optimality conditions for nonsmooth locally Lipchitz problems have been given in terms of the Clarke generalized subdifferential [4, 5, 18, 19, 20, 21].

Brando et al. [4, 5] weekend convexity requirement to invexity and obtained Karush-Kuhn optimality conditions for a class of nonsmooth continuous time Lipchitz programming. We begin with an optimality condition for a single continuous-time problem.

Lemma 1 *[4]. Let $\bar{x}$ be a feasible solution of (CT) and $f(t,.)$ and $g_i(t,.), i \in M$ be Lispchitz near $\bar{x}(t)$. Then there exist $\mu_0 \in \mathbb{R}, \mu_i \in L_\infty^m[0,T], i \in M$, such that*

$$0 \in \int_0^T [\mu_0 \partial_c f(t, \bar{x}(t)) + \sum_{i \in M} \mu_i(t) \partial_c g_i(t, \bar{x}(t))]dt,$$

$$(\mu_0, \mu(t)) = (\mu_0, \mu_1(t), \ldots, \mu_m(t)) \neq 0, \quad a.e. \quad t \in [0,T],$$

$$\mu_i(t)g_i(t, \bar{x}(t)) = 0, \mu_0 \geq 0, \mu_i(t) \geq 0, \quad a.e. \quad t \in [0,T], \quad i \in M.$$

Lemma 2 *[4] Let $\bar{x}$ be a feasible solution of (CT). Suppose that $f(t,.)$ and $g_i(t,.), i \in M$ are invex at $\bar{x}(t)$, with the same $\eta(x(t), \bar{x}(t))$ for all functions. Suppose, further that there exists $\bar{\lambda} \in L_\infty^m[0,T]$ such that*

$$0 \leq \int_0^T \{f^0(t, \bar{x}(t); h(t)) + \sum_{i=1}^m \bar{\lambda}_i(t)g_i^0(t, \bar{x}(t); h(t))\}dt,$$

$$\bar{\lambda}_i(t)g_i(t, \bar{x}(t)) = 0, \bar{\lambda}_i(t) \geq 0 \ a.e. \ t \in [0,T], \ i \in M, \forall h \in L_\infty^n[0,T],$$

then $\bar{x}$ is a global optimal solution of (CT).

It is usual to assume some kind of regularity condition on the restrictions of the problem to make sure that Lagrange multiplier associated with objective function will be nonzero. These regularity conditions are usually referred to as constraint qualifications. Consider the following constraint qualification(CQ) from [4],

$$(CQ) \quad \cap_{i \in M} K(g_i, \bar{x}) \neq \theta,$$

where

$$K(g_i, \bar{x}) = \{h \in L_\infty^n[0,T] : g_i^0(t, \bar{x}(t); h(t)) < 0, \quad a.e. \ t \in A_i(\bar{x}), i \in M\}$$

and $A_i(\bar{x}) = \{t \in [0,T] : g_i(t, \bar{x}(t)) = 0\}$.

In Lemma 1, if (CQ) holds, then the Lagrange multiplier associated with the objective function is nonzero. Then we obtain the following Kuhn-Tucker necessary type condition.

Lemma 3 *[4]. Let $\bar{x}$ be a feasible solution of (CT) and the constraint qualification (CQ) is satisfied. Assume that $f(t,.)$ and $g_i(t,.), i \in M$ are Lispchitz near $\bar{x}(t)$. Then there exist $\mu_i \in L_\infty^m[0,T], i \in M$, such that*

$$0 \in \int_0^T [\partial_c f(t, \bar{x}(t)) + \sum_{i \in M} \mu_i(t) \partial_c g_i(t, \bar{x}(t))] dt,$$

$$\mu_i(t) g_i(t, \bar{x}(t)) = 0, \quad \mu_i(t) \geq 0, \quad a.e. \quad t \in [0,T], \quad i \in M.$$

We discuss optimality conditions for (MCTP) under various generalized invexity conditions. We begin with an optimality condition for a single continuous-time problem. We consider the following continuous-time problem:

$$P_k(x^*) \quad \min \quad \int_0^T f_k(t, x(t)) dt$$

$$s.t. \quad g_i(t, x(t)) \leq 0 \quad i \in M \quad a.e. \quad t \in [0,T]$$

$$\int_0^T f_j(t, x(t)) dt \leq \int_0^T f_j(t, x^*(t)) dt.$$

The following lemma from [8] connects the efficient solutions of (MCTP) and $P_k(.)$.

Lemma 4 *[8]. A point $x^* \in F_P$ is an efficient solution, for (MCTP) if and only if x^* solves $P_k(x^*)$ for all $k = 1, 2, \ldots, r$.*

Now we give necessary and sufficient optimality conditions theorems for (MCTP).

Theorem 4 *(Necessity) If $\bar{x}$ is an efficient solution of (MCTP), and $P_k(\bar{x})$ satisfies the constraint qualification (CQ) at $\bar{x}$ for some k, then there exist piecewise smooth function $\tau^0 \in \mathbb{R}^r$, and $\lambda^0 : I \to R^m$ such that,*

$$0 \in \int_0^T (\sum_{i=1}^r \tau_i^0 \partial_c f_i(t, \bar{x}(t)) + \sum_{j \in M} \lambda_j^0(t) \partial_c g_j(t, \bar{x}(t)) dt \quad (1)$$

$$\lambda_j^0(t) g_j(t, \bar{x}(t)) = 0, \lambda_j^0(t) \geq 0, j \in M, \ a.e. \ t \in [0,T], \ \sum_{i=1}^r \tau_i^0 = 1, \tau_i^0 \geq 0. \quad (2)$$

Proof. Let $\bar{x}$ be an efficient solution of (MCTP) then by Lemma 4, $\bar{x}$ is an optimal solution of $P_k(\bar{x})$. Therefore by Lemma 1 there exist nonnegative $\tau \in \mathbb{R}^{r-1}$ and $\lambda \in L_\infty^m[0,T]$, such that

$$0 \in \int_0^T \{\partial_c f_k(t, \bar{x}(t)) + \sum_{i \in L, i \neq k} \tau_i \partial_c f_i(t, \bar{x}(t)) + \sum_{j \in M} \lambda_j(t) \partial_c g_j(t, \bar{x}(t))\} dt,$$

$$\lambda_j(t) g_j(t, \bar{x}(t)) = 0, \lambda_j(t) \geq 0, j \in M, \tau_i \geq 0, i \in L \quad a.e. \quad t \in [0,T].$$

Now let

$$\tau_k^0 = \frac{1}{1 + \sum_{i=1 i \neq k}^r \tau_i}, \quad \tau_i^0 = \frac{\tau_i}{1 + \sum_{i=1 i \neq k}^r \tau_i}, \quad i \in L, i \neq k,$$

$$\lambda_j^0(t) = \frac{\lambda_j(t)}{1 + \sum_{i=1 i \neq k}^r \tau_i}, \quad j = 1, 2, \ldots, m.$$

Then conditions (1) and (2) hold. $\qquad\qquad\qquad\qquad\qquad\qquad\qquad\qquad\qquad\qquad\qquad$ □

Theorem 5 *(Sufficiency) Suppose that there exist a feasible solution x^* for (MCTP), $\tau \in \mathbb{R}^r$ and $\lambda \in L_\infty^m[0, T]$ such that:*

$$0 \in \int_0^T \left(\sum_{i=1}^r \tau_i \partial_c f_i(t, x^*(t)) + \sum_{j \in M} \lambda_j(t) \partial_c g_j(t, x^*(t)) dt, \right. \tag{3}$$

$$\lambda_j(t) g_j(t, x^*(t)) = 0, \quad \lambda_j(t) \geq 0, j \in M, \quad \sum_{i=1}^r \tau_i = 1, \quad \tau_i \geq 0. \tag{4}$$

If $f_i(t, .), i \in L$, are strictly invex and $g_j(t, .), j \in M$ are invex at x^* throughout $[0, T]$, with the same $\eta(x(t), x^*(t))$ for all functions, then x^* is an efficient solution for (MCTP).

Proof. Suppose that x^* is not an efficient solution for (MCTP), then there exists $x \in F_P$, such that

$$\int_0^T f(t, x(t)) dt \leq \int_0^T f(t, x^*(t)) dt.$$

By assumptions on $f_i, i \in L$ and $g_j, j \in M$, we have

$$\int_0^T \{\lambda_j(t) g_j^0(t, x^*(t); \eta(x(t), x^*(t)))\} dt \leq 0, \tag{5}$$

$$\int_0^T f_i^0(t, x^*(t); \eta(x(t), x^*(t))) dt < 0. \tag{6}$$

Now from (5) and (6) it follows that

$$\int_0^T \sum_{i=1}^r \tau_i f_i^0(t, x^*(t); \eta(x(t), x^*(t))) +$$

$$\sum_{j=1}^m \lambda_j(t) g_j^0(t, x^*(t); \eta(x(t), x^*(t))) dt < 0,$$

which contradicts (3). Therefore we conclude that x^* is an efficient solution of (MCTP). □

Remark 1 *Theorem 5 also holds under any of the following different types of assumptions:*

(i) If $f_i(t, .), i \in L$, are QIX and $\lambda_j(t)g_j(t, .), j \in M$ are SQIX at x^* throughout $[0, T]$, with the same $\eta(x(t), x^*(t))$ for all functions.

(ii) If $f_i(t, .), i \in L$, are SQIX and $g_j(t, .), j \in M$ are QIX at x^* throughout $[0, T]$, with the same $\eta(x(t), x^*(t))$ for all functions.

Remark 2 *Consider the following single-objective problem;*

$$SCP(x^*) \quad \min \quad \sum_{i \in L} \int_0^T f_i(t, x(t))dt$$

$$s.t. \quad g_i(t, x(t)) \leq 0 \ i \in M, \ a.e. \quad t \in [0, T],$$

$$\int_0^T f_j(t, x(t))dt \leq \int_0^T f_j(t, x^*(t))dt, \quad \forall j \in L.$$

Now, the necessary optimality condition for problem (MCTP) can be proved via the scalerization $(SCP(x^*))$ and the following lemma which can be verified easily.

Lemma 5 *A feasible solution x^* of (MCTP) is an efficient solution for (MCTP) if and only if x^* is an optimal solution of the single-objective problem $(SCP(x^*))$.*

We now give an example to verify the necessary optimality condition for a particular (MCTP).

Examples 2 *Consider the following multiobjective continuous-time problem:*

$$(MCP) \min \quad [\int_0^1 f_1(t, x(t))dt, \int_0^1 f_2(t, x(t))dt]$$

$$s.t. \quad g_i(t, x(t)) \leq 0, i \in M = \{1, 2, 3\}, x \in X, a.e. \quad t \in [0, 1]$$

where
$$f_1(t, x(t)) := |x_1(t) - t|, \quad f_2(t, x(t)) := |x_2(t) - t|,$$

$$g_1(t, x(t)) := -x_1(t), g_2(t, x(t)) := -x_2(t), \quad g_3(t, x(t)) = x_1(t) + x_2(t) - 2t,$$

and $x : [0, 1] \rightarrow \mathbb{R}^2$ is defined by $x(t) = (x_1t, x_2t), x_1, x_2 \in \mathbb{R}$.

It can be easily verified that $x^*(t) = (t, t)$ is an efficient point of the problem (MCTP) and $P_K(x^*)$ for $k = 1, 2$ satisfy constraint qualification. The necessary optimality condition is satisfied for $\tau_1 = \frac{1}{2}, \tau_2 = \frac{1}{2}, \lambda_1(t) = 0, \lambda_2(t) = 0, \lambda_3(t) = t$.

Remark 3 *In the case where we have continuous differentiability of the functions involved, the Karush-Kuhn-Tucker necessary condition reduce to the following form ;(see [32]),*

$$\int_0^T (\sum_{i=1}^r \tau_i^0 \nabla f_i(t, \bar{x}(t)) + \sum_{j \in M} \lambda_j^0(t) \nabla g_j(t, \bar{x}(t))) h(t) dt = 0, \forall h \in L_\infty^n[0, T], \qquad (7)$$

$$\sum_{j=1}^m \lambda_j^0(t) g_j(t, \bar{x}(t)) = 0, \lambda_j^0(t) \geq 0, \ a.e. \ t \in [0, T], \ \sum_{i=1}^r \tau_i^0 = 1, \tau_i^0 \geq 0. \qquad (8)$$

4 Duality for nonsmooth multiobjective continuous time problems

Several approaches to duality for the continuous-time problems may be found in the literature. These include the use of Karush-Kuhn-Tucker optimality conditions to establish some duality theorems. A vast number of works have appeared dealing with duality in multiobjective programs under different assumptions of convexity, for example, Zalmai [31] established the duality results for continuous time homogenous programming under convexity and differentiability assumptions.

Zalmai [31] considered a class of differentiable continuous-time programming problems of the form;

$$(\bar{P}) \quad \min \quad \int_0^T f(x(t), t) dt$$

$$s.t. \quad g_i(x(t), t) + b(t) \leq 0 \quad i \in M \quad a.e. \quad t \in [0, T], \quad x \in X,$$

and presented Mond-Wier dual given below and established various duality results, weak, strong and converse duality theorems under convexity assumptions.

The dual problem corresponding to $(\bar{P})$ is the following;

$$\max \phi(x, u) = \int_0^T \{f(x(t), t) + u(t)[g(x(t), t) + b(t)]$$
$$- [\nabla f(x(t), t) + \sum_{i=1}^m u_i(t) \nabla g(x(t), t)] x(t)\} dt$$
$$s.t.:$$

$$\int_0^T [\nabla f(x(t), t) + \sum_{i=1}^m u_i(t) \nabla g(x(t), t)] y(t) dt \geq 0, \forall y \in L_\infty^n[0, T], u(t) \geq 0, \ a.e. \ t \in [0, T]$$

where $x \in X$ and $u \in L_\infty^n[0, T]$.

In the following two dual type models for (VCTP) are proposed and duality relationships are established under invexity assumptions. Let e be the vector in $\mathbb{R}^m$ whose components are all one.

Wolf Continuous-Time Dual(WCTD).

$$\max \ \int_0^T (f(t, u(t)) + w(t)'g(t, u(t))e)dt$$
$$\text{s.t.:}$$

$$0 \in \int_0^T \{\sum_{i=1}^r \tau_i \partial_c f_i(t, u(t)) + \sum_{i=1}^m w_i(t)' \partial_c g_i(t, u(t))\} dt, \tag{9}$$

$$\sum_{i=1}^r \tau_i = 1; \tau_i \geq 0, \ w_j(t) \geq 0, \ j \in M \ a.e. \ t \in [0, T], u \in X. \tag{10}$$

Mond-Weir Continuous-Time Dual (MCTD)

$$\max \ \int_0^T f(t, u(t))dt$$
$$\text{s.t.:}$$

$$0 \in \int_0^T \{\sum_{i=1}^r \tau_i \partial_c f_i(t, u(t)) + \sum_{j=1}^m w_j(t)' \partial_c g_j(t, u(t))\} dt, \tag{11}$$

$$w_j(t)g_j(t, u(t)) \geq 0, j \in M, \tag{12}$$

$$\sum_{i=1}^r \tau_i = 1, \tau_i \geq 0, i \in L, \ w_j(t) \geq 0, j \in M, \ a.e. \ t \in [0, T], u \in X$$

We now establish weak duality and strong duality results between the primal problem (VCTP) and its dual (WCTD).

Theorem 6 *(Weak Duality) Let x be feasible for* (VCTP) *and* (u, τ, w) *be feasible for* (WCTD). *If any of the following condition holds:*

(a) *$f_i(t, .), i \in L$ and $g_i(t, .), i \in M$ are invex at u throughout $[0, T]$, with respect to η and $\tau_i > 0$,*

(b) *$f_i(t, .), i \in L$ are strictly invex and $g_j(t, .), j \in M$ are invex at u throughout $[0, T]$ with respect to η.*

Then the following cannot hold:

$$\int_0^T f(t, x(t))dt \leq \int_0^T \{f(t, u(t)) + w(t)'g(t, u(t))e\}dt, \tag{13}$$

Proof. Suppose contrary to the result of theorem that (13) holds. Since $\tau_i > 0$, we obtain

$$\int_0^T \{\sum_{i \in L} \tau_i(f_i(t, x(t)) - (f_i(t, u(t)) + w(t)'g(t, u(t)))\}dt < 0. \tag{14}$$

If (a) holds, by the inequality (14) we have

$$\int_0^T [\sum_{i=1}^r \tau_i f_i^0(t, u(t)); \eta(x(t), u(t))$$

$$+ \sum_{i=1}^m w_i(t)'g_i{}^0(t, u(t)); \eta(x(t), u(t))]dt < 0.$$

This contradicts the inclusion (9). The rest of the proof is similar to the first part. Hence the proof is complete.

□

Corollary 1 Assume that weak duality Theorem 6 holds between (VCTP) and (WCTD). If u^0 is feasible for (VCTP), and (u^0, τ^0, ω^0) is feasible for (WCTD) with $w^0(t)'g(t, u^0(t)) = 0$. Then u^0 is efficient for (VCTP) and (u^0, τ^0, ω^0) is efficient for (WCTD).

Proof. Suppose u^0 is not efficient for (VCTP). Then there exists a feasible solution x for (VCTP) such that

$$\int_0^T f(t, x(t))dt \leq \int_0^T f(t, u^0(t))dt.$$

Since $w^0(t)'g(t, u^0(t)) = 0$ we obtain

$$\int_0^T f(t, x(t))dt \leq \int_0^T \{f(t, u^0(t))dt + w^0(t)'g(t, u^0(t))e\}dt.$$

This contradicts the weak duality theorem. Hence u^0 is efficient for (VCTP). Now suppose that (u^0, τ^0, ω^0) is not efficient for (WCTD). Then there exists (u, τ, ω) feasible for (WCTD) such that

$$\int_0^T \{f(t, u(t)) + w(t)'g(t, u(t))e\}dt \geq \int_0^T \{f(t, u^0(t)) + w^0(t)'g(t, u^0)e\}dt.$$

Since $w^0(t)'g(t, u^0(t)) = 0$, then

$$\int_0^T \{f(t, u(t))dt + w(t)'g(t, u(t))edt \geq \int_0^T f(t, u^0(t))dt.$$

This contradicts weak duality. Hence (u^0, τ^0, ω^0) is efficient for (WCTD).

□

Theorem 7 *(Strong Duality) Let u^0 be efficient for (VCTP) and assume that u^0 satisfies the constraint qualification(CQ) for $P_k(u^0)$ for at least one $k \in L$. Then, there exist $\tau^0 \in \mathbb{R}^r$ and piecewise smooth $w^0 : I \to \mathbb{R}^p$ such that (u^0, τ^0, w^0) is feasible for (WCTD) and $w^{0'}(t)'g(t, u^0(t)) = 0$. If weak duality Theorem 6 also holds between (VCTP) and (WCTD), then (u^0, τ^0, w^0) is efficient for (WCTD).*

Proof. Since u^0 satisfies the (CQ) for at least one k, it follows from Theorem 4 that there exist $\tau^0 \in \mathbb{R}^r$ and piecewise smooth $w^0 : I \to \mathbb{R}^m$ such that;

$$0 \in \int_0^T \{\sum_{i=1}^r \tau_i^0 (\partial_c f_i(t, u^0(t)) + w^{0'}(t)\partial_c g(t, u^0(t)))\}dt, \tag{15}$$

$$\sum_{i=1}^r \tau_i^0 = 1, \tau_i^0 \geq 0,$$

$$w^{0'}(t)g(t, u^0(t)) = 0, w^0(t) \geq 0, i \in L, a.e.\ t \in [0, T].$$

Now it follows from (15), that (u^0, τ^0, w^0) is feasible for (WCTD). Also $w^{0'}(t)g(t, u^0(t)) = 0$ and weak duality holds between (VCTP) and (WCTD). The result follows from Corollary 3.1. $\qquad\qquad\qquad\qquad\qquad\qquad\qquad\qquad\qquad\qquad\qquad\qquad\qquad\qquad\square$

Duality between (VCTP) and (MCTD)

We now establish weak duality and strong duality results between the primal problem (VCTP) and its dual (MCTD).

Theorem 8 *(Weak Duality) Let x be feasible for (VCTP) and (u, τ, w) be feasible for (MCTD). If*

(i) $f_i(t, .), i \in L$ are strictly invex at u throughout $[0, T]$ with respect to η,

(ii) $g_i(t, .), i \in M$ are invex at u throughout $[0, T]$ with respect to η,

then the following cannot hold:

$$\int_0^T f(t, x(t))dt \leq \int_0^T f(t, u(t))dt. \tag{16}$$

Remark 4 *The weak duality theorem also holds under the following different types of assumptions;*

(a) $f_i(t, .), i \in L$ are SQIX at u throughout $[0, T]$ with respect to η, and $g_i(t, .), i \in M$ is QIX at u throughout $[0, T]$ with respect to the same η.

(b) $\tau_i > 0, f_i(t,.), i \in L$ *are QIX at u throughout* $[0,T]$ *with respect to* η, *and*
$\omega_i(t)g_i(t,.), i \in M$ *are SQIX at u throughout* $[0,T]$ *with respect to the same* η.

Corollary 2 *Assume that weak duality Theorem 6 holds between (VCTP) and (MCTD). If* u^0 *is feasible for (VCTP), and* (u^0, τ^0, ω^0) *is feasible for (MCTD). Then* u^0 *is efficient for (VCTP) and* (u^0, τ^0, ω^0) *is efficient for (MCTD).*

Theorem 9 *(Strong Duality)Let* u^0 *be efficient for (VCTP) and assume that* u^0 *satisfies the constraint qualification(CQ) for* $P_k(u^0)$ *for at least one* $k \in L$. *Then there exist* $\tau^0 \in \mathbb{R}^r$ *and piecewise smooth* $w^0 : I \to \mathbb{R}^m$ *such that* (u^0, τ^0, w^0) *is feasible for (MCTD). If weak duality theorem also holds between (VCTP) and (MCTD) then* (u^0, τ^0, w^0) *is efficient for (MCTD).*

Remark 5 *In the case where we have continuous differentiability of the functions involved, then the Wolf dual type is equivalent to the following problem [33]:*

$$\max \int_0^T \{(f(t,u(t)) + w'(t)g(t,u(t))e\}dt,$$

subject to

$$\int_0^T ((\sum_{i=1}^r \tau_i \nabla f_i(t,\bar{x}(t)) + \sum_{j \in M} w_j(t)\nabla g_j(t,\bar{x}(t)))h(t)dt = 0, \forall h \in L_\infty^n[0,T], \qquad (17)$$

$$\sum_{i=1}^r \tau_i = 1, \tau_i \geq 0, \ w_j(t) \geq 0, \quad j \in M \ a.e. \ t \in [0,T]. \qquad (18)$$

References

[1] J. Abrham and R.N. Buir, Kuhn-Tucker conditions and duality in continuous programming. Utilitas Mathematica 16 (1979) 15–37.

[2] K.J. Arrow, and M.D. Intriligator, Editors, Handbook of Mathematical Economics, North Holland, Amsterdam, Holland, Vol. 3, 1986.

[3] R. Bellman, Bottleneck problems and dynamic programming, Proceedings of the National Academy of Sciences of USA, 39 (1953) 947–951.

[4] A.J.V. Brandao, M.A. Rojas-Medar, and G.N. Silva, Nonsmooth Continuous-Time Optimization Problems: Necessary Conditions, Computers and Mathematical with Applications 41 (2001) 1477–1486.

[5] A.J.V. Brandao, M.A. Rojas-Medar, and G.N. Silva, Nonsmooth Continuous-Time Optimization Problems: Sufficient Conditions, Journal of Mathematical Analysis and Applications 227 (1998) 305–318.

[6] F.H. Clarke, Optimization and Nonsmooth Analysis, Wiley-Interscience, New York, 1983.

[7] F.H. Clarke, Yu.S. Ledyaev, R.J. Stern, and P. Wolenski, Nonsmooth Analysis and Control Theory, 178 Springer-Verlag, Graduate Texts in Mathematics, New York, 1998.

[8] V. Chankong, and Y. Haimes, Multiobjective Decision Making; Theory and Methodology, North-Holland, New York, 1983.

[9] B.D. Craven, Invex functions and constraint local minima, Journal of Australian Mathematical Society, Series B 24 (1981) 357–366.

[10] W.H. Farr and M.A. Hanson, Continuous time programming with nonlinear timedelayed constraints, Journal of Mathematical Analysis and Applications 46 (1974) 41–61.

[11] W.H. Farr and M.A. Hanson, Continuous time programming with nonlinear constraints, Journal of Mathematical Analysis and Applications 45 (1974) 96–115.

[12] R. Grinold, Symmetric duality for continuous linear programs, SIAM Journal on Applied Mathematics 18 (1970) 84–97.

[13] M. A. Hanson, On sufficiency of the Kuhn-Tucker conditions, Journal of Mathematical Analysis and Applications 80 (1981) 545–550.

[14] M.A. Hanson, and B. Mond, A class of Continuous convex Programmingproblems, Journal of Mathematical Analysis and Applications 22 (1968) 427–437.

[15] N. Levinson, A class of Continuous linear Programming problems, Journal of Mathematical Analysis and Applications 16 (1966) 73–83.

[16] C. Lee, and H.C Lai, Parameter-free dual models for fractional programming with generalized invexity, Annals of Operations Research 133 (2005) 47–61.

[17] S.M.N. Islam, and B.D. Craven, Computation of Nonlinear Continuous Optimal Growth Models; Experiments with Optimal Control Algorithms and Computer Program, Economic Modeling, Vol. 18 (2002) 551–568.

[18] S. Nobakhtian and M.R. Pouryayevali, Optimality Conditions and Duality for Nonsmooth Fractional Continuous-Time Problems, to appear in Journal of Optimization Theory and Applications.

[19] S. Nobakhtian, and M.R. Pouryayevali, Optimality criteria for nonsmooth continuous-time Problems of multiobjective optimization, Journal of Optimization Theory and Applications 136 (2008) 69–76.

[20] S. Nobakhtian, and M.R. Pouryayevali, Duality for nonsmooth continuous-time problem, Journal of Optimization Theory and Applications (2008) 136 77–85.

[21] S. Nobakhtian, Nonsmooth multiobjective continuous-time problems with generalized invexity, Journal of Global Optimization 43, (2009) 593–606.

[22] T.W. Reiland and M.A. Hanson,, Continuous time programming with nonlinear time delayed constraints, Journal of Mathematical Analysis and Applications 46 (1974) 41–61.

[23] T.W. Reiland and M.A. Hanson, Generalized Kuhn-Tucker conditions and duality for continuous nonlinear programming problems, Journal of Mathematical Analysis and Applications 74 (1980) 578–598.

[24] T.W. Reiland, Optimality and duality in continuous programming. I. Convex programs and a theorem of the alternative, Journal of Mathematical Analysis and Applications 77 (1980) 297–325.

[25] A. Schy, and D.P. Giesy, Multiobjective Optimization Techniques for Design of Aircraft Control Systems, Multiobjective Optimization in Engineering and in the Science, Edited by W. Stadler, Plenum Press, New York, (1988) 225–262.

[26] W. Stadler, Multiobjective Optimization in Mechanics: A survey, Applied Mechanics Reviews. Vol. 37, (1984) 277–286.

[27] W. Stadler, Multiobjective Optimization in Engineering and in the Science, Plenum Press, New York, 1988.

[28] G.J. Zalmai, Continuous-Time Multiobjective Fractional Programming, Optimization, Vol. 37 (1996) 1–25.

[29] G.J. Zalmai, Proper Efficiency Conditions and Duality Models for a Class of Nonsmooth Continuous-Time Multiobjective Fractional Programming Problems, Southeast Asian Bulletin of Mathematics, Vol. 27, (2003) 155–186.

[30] G.J. Zalmai, A continuous-time generalization of Gordans transposition theorem, Journal of Mathematical Analysis and Applications 110 (1985) 130–140.

[31] G.J. Zalmai, Duality in Continuous-Time Homogeneous Programming, Journal of Mathematical Analysis and Applications 111 (1985) 433–448.

[32] G.J. Zalmai, The Fritz John and Kuhn- Tucker Optimality Conditions in Continuous-Time Programming, Journal of Mathematical Analysis and Applications, 110 (1985) 503–518.

[33] G.J. Zalmai, Optimality Conditions and Lagrangian Duality in Continuous-time Non-linear Programming, Journal of Mathematical Analysis and Applications 109 (1985) 426–452.

[34] W.I. Zangwill, Nonlinear Programming: A Unified Approach, Prentice-Hall, Englewood Cliffs, N.J.,1967.

Index

www.ingramcontent.com/pod-product-compliance
Lightning Source LLC
Chambersburg PA
CBHW080020240326
41598CB00075B/513